T0142222

Springer Series on Touch and Haptic Systems

More information about this series at http://www.springer.com/series/8786

Philipp Beckerle

Human-Robot Body Experience

Springer

Philipp Beckerle
Chair of Autonomous Systems and Mechatronics
Department of Electrical Engineering
University of Erlangen-Nuremberg
Erlangen, Germany

ISSN 2192-2977 ISSN 2192-2985 (electronic)
Springer Series on Touch and Haptic Systems
ISBN 978-3-030-38690-0 ISBN 978-3-030-38688-7 (eBook)
https://doi.org/10.1007/978-3-030-38688-7

This Springer imprint is published by the registered company Springer Nature Switzerland AG
The registered company address is: Gewerbestrasse 11, 6330 Cham, Switzerland

Series Editors' Foreword

This is the eighteenth volume of *Springer Series on Touch and Haptic Systems*, which is published as a collaboration between **Springer** and the **EuroHaptics Society**.

Human-Robot Body Experience describes a set of human-in-the-loop experiments based on technologies in haptics, robotics, and virtual reality. It highlights the paramount importance of haptics in human-robot interactions. Overall, the results show a significantly stronger embodiment experience when haptic feedback is provided.

Robotic hand and leg illusions are discussed in depth in order to analyze haptic mechanisms. Several experiments that include interaction with vibrotactile and force feedback are also presented in terms of the light they shed on human-robot interconnection.

The book includes eight chapters grouped into four parts. The first part introduces the fundamentals of human-robot interaction. The second and third parts are focused on the upper and lower robot limbs, respectively, and the fourth part defines some guidelines on their design and future lines of research.

This volume reviews many important issues in human-robot interaction and paves the way for future research work as well as providing recommendations for the design of more effective human-robot interfaces.

Madrid, Spain
Ulm, Germany
Birmingham, UK
January 2021

Manuel Ferre
Marc O. Ernst
Alan Wing

Foreword

The body experience of users of assistive robots and other applications with an intensive mix of cognitive and physical human-robot interactions is a very challenging field of fundamental research with highly promising prospects for future technical applications, especially in haptics. Providing users with intuitive information, haptic feedback can enrich the control possibilities given to the user and thereby increase the usability of assistive systems, e.g., in teleoperation or prosthetics. With a very holistic approach, this monograph of Prof. Philipp Beckerle considers this fascinating research challenge from the perspectives of diverse disciplines in human and engineering sciences. In a very tangible fashion, Philipp brings together these perspectives outlining the enormous potential and impact of human-in-the-loop experiments for our understanding of human body experience and the design of haptic feedback devices and assistive controllers. This monograph will empower readers from different fields to understand the human-robot body experience, as Philipp coins it, in a very comprehensive way. The interpretation of the human-robot interaction that Philipp introduces in this book will certainly attract the interest of many researchers in our community.

Bridging between the disciplines, the first part of the book effectively provides readers with the necessary fundamentals, presents existing experimental designs, and their requirements. From this part, readers from different fields are wonderfully picked up and brought to a common understanding of the interrelations of human body experience and robotic assistance. In the second and third parts, Philipp's excellent monograph brings those fields together by suggesting and discussing human-in-the-loop experimental designs to probe human-robot body experience regarding the upper and lower limbs. The presented approaches and studies nicely outline the influence of haptics and control and how experience-related aspects could be considered in their design. Concluding the monograph, the fourth part discusses exciting technical solutions and provides a roadmap for future research on bidirectional human-machine interfaces and non-functional haptic feedback. Providing a very comprehensive picture of the scientific challenge and highlighting promising future research directions, I expect this monograph to become a staple reference in this field.

Siena, Italy
February 2021

Domenico Prattichizzo

Preface

Robots and related assistive technologies are entering our daily lives and starting to share our professional and private workspaces. During tight interaction, the way of how we experience our bodies is influenced by interaction with robots and assistive devices, e.g., a prosthesis might feel to be part of the user's body. This monograph gives a broad overview of research of the body experience of human individuals who are directly interacting with robots and assistive devices. It presents results from the author's interdisciplinary research at the intersection of psychology, cognitive and computer science, neuroscience, and engineering. Going beyond disciplinary boundaries, human-in-the-loop experiments based on psychological paradigms are suggested to empirically evaluate how users experience devices and systematically analyze how body experience influences system, interface, and control design.

Personally, I would like to thank all collaborators, supporters, and students who put their hard work and dedication to the research condensed in this monograph. Special thanks go to the faculties of Electrical Engineering and Information Technology at TU Dortmund and Mechanical Engineering at TU Darmstadt as well as SIRS lab at the University of Siena for hosting and facilitating my research. Moreover, I highly appreciate the support provided by the IEEE Technical Committee on Haptics and acknowledge the support from the German Research Foundation (DFG) through the projects "Users Body Experience and Human-Machine Interfaces in (Assistive) Robotics" (no. BE 5729/3&11).

I am fully convinced that we will only be able to understand and shape human-robot body experience through intensive exchange and open discourse within and across disciplines. Numerous discussions with colleagues have shown me that despite recent insights, there is a long way ahead of us. Besides capturing the state of the art and providing concrete experimental approaches, this monograph discusses future directions to hopefully provide guidance to this endeavor.

Darmstadt, Germany
September 2020

Philipp Beckerle

Contents

Part I
Fundamentals and Requirements

Chapter 1
Introduction

Abstract Through its flexibility, human body experience can integrate technical artifacts. This includes not only passive tools but also advanced robotic devices and is, especially, promising for situations with tight human-robot interaction. As the interaction of humans and robots is recently getting closer, e.g., in collaborative working scenarios or healthcare applications, an improved understanding of how devices are embodied by their users is of immense societal potential. To explore how human body experience and human-robot interaction relate to each other, fundamental aspects of body experience and representations need to be understood and this knowledge has to be made accessible to engineering design. This chapter gives a detailed motivation of why the consideration of human body experience is an important aspect of the evaluation of human-robot interaction and human-machine interfaces. A concise description of the objectives of this monograph outlines the need for an interdisciplinary approach and explains the focus on human-in-the-loop experiments and haptic interfaces. Finally, the structure of the monograph and the content of the individual parts and chapters are presented.

1.1 Motivation

Human body experience is remarkably flexible and can integrate artifacts ranging from simple passive tools like a hammer to advanced robotic devices such as prosthetic hands [1, 2]. With the recent rise of wearable robotics, the body experience of human individuals using such devices is of paramount importance and high interest for scientific research but also to the aging society. To understand fundamentals of human experience, the influence of the robot, and the interaction of both partners, i.e., both agents, the combination of perspectives, knowledge, and methods from various disciplines spanning from psychology, cognitive science, and neuroscience to computer science and engineering is required.

In the past decades, psychological research developed experimental paradigms to explore how foreign objects are perceived as a part of one's body, e.g., the rubber hand illusion [3]. In the rubber hand illusion experiment, participants experience a rubber hand as their own one due to targeted multisensory stimulation (see Chap. 2

P. Beckerle, *Human-Robot Body Experience*, Springer Series on Touch and Haptic Systems, https://doi.org/10.1007/978-3-030-38688-7_1

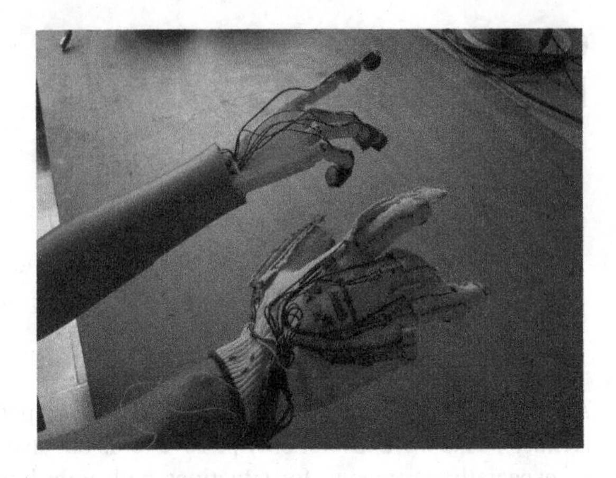

Fig. 1.1 A human-in-the-loop experimental setup: the user controls the robotic hand via a sensory glove and perceives vibrotactile feedback at the fingertips

for details). The applicability of such bodily illusion paradigms as means for the technical design of (assistive) robots is subject to very active research [1, 4–6]. The observed embodiment of artifacts is due to crossmodal integration of various sensory information such as vision, touch, and proprioception [7]. This multisensory integration might be influenced by technical variations of design, control, and feedback [5]. Unraveling the interrelations of technical design and its influence on fundamental experience processes might be key to implement devices that are integrated and accepted by their users [1, 4, 5, 8].

Despite these promising prospects, the human body experience during tight interaction with robots is a rather new and very challenging field. Besides missing knowledge regarding the fundamentals of human body experience itself, the integration of robotic devices in psychological experiments demands additional efforts, e.g., adding sensors, actuators, and control hardware (see Fig. 1.1). By altering specific technical factors during interaction of human and robot, such human-in-the-loop experiments help to understand the embodiment of robotic devices [5, 9, 10]. Yet, they currently rely on proprietarily designed experimental setups and specific requirements and guidelines seem necessary for structured design [10]. Appropriate human-in-the-loop experiments can be expected to improve the evaluation of factors supporting and affecting bodily illusions and could thereby help to develop a novel generation of human-machine interfaces that improve the embodiment of robotic devices practically [1, 4, 5]. In the contexts of assistive robots, e.g., wearable robotics, and other scenarios with tight human-robot interaction like teleoperation, haptics are of particular importance due to the physical contact between human and robot [5, 8, 11].

1.2 Objectives and Approach

This monograph presents recent research of the body experience of human individuals focusing on scenarios where they are using robotic devices such as assistive robots or teleoperation systems, which is referred to as human-robot body experience. By putting various previous works[1] into a broader context, this monograph systematically evaluates how body experience influences system and control design and, particularly, (haptic) interface development. The focus of the monograph is on human-in-the-loop experiments that help to empirically evaluate how users experience devices while accounting for multisensory influences. Besides reporting and discussing psychological examinations, the influence of varying aspects of engineering design is investigated, e.g., variations of haptic interfaces or robot control. As haptics are of paramount importance in tight human-robot interaction, it is explored with respect to modality as well as temporal effects. Moreover, potential applications of cognitive models of human body experience in robotics are discussed and a Bayesian modeling approach is presented. Results from expert studies and human-in-the-loop experiments point out specific influences of design and control, from which design considerations are concluded and directions for future research and development are suggested.

1.3 Structure

The first part of the monograph introduces and motivates the topic, explains fundamental terms, and gives an in-depth analysis of the experimental requirements. Furthermore, Chap. 2 introduces human-in-the-loop experimental paradigms and their perspective use in human-robot interaction research. Based on the literature and experimental results, the requirements to be considered when designing robotic devices to explore human body experience are analyzed and generalized.

Focusing on the upper limbs, the second part of the monograph revolves around the rubber hand illusion paradigm. Chapter 3 discusses the experimental merits of introducing robotic devices and tactile feedback. Additionally, it investigates the influence of motion and feedback delays to derive design considerations. In Chap. 4, virtual hand illusion experiments are analyzed to provide deeper insights with respect to feedback modalities and the influence of semi-autonomous control.

Covering lower limb body experience, the third part of the monograph relies on rubber foot and robotic leg illusion studies. Chapter 5 presents how to tackle the technical challenges of implementing human-in-the-loop experiments for lower limb bodily illusions and confirms the transferability of the rubber hand illusion concept. In Chap. 6, the potential of cognitive models of body experience in robotics are discussed and a Bayesian approach to model bodily illusions is presented and discussed.

The fourth part of the monograph discusses design considerations and derives directions to guide future research. Chapter 7 discusses design and implementation approaches with a focus on the advancement of experimental designs using wireless sensor gloves, integrated psychophysiological measurement, and new hand/arm

[1] Authored or co-authored by the author of this monograph.

design concepts. Finally, Chap. 8 discusses bidirectional human-machine interfaces, which appear to be a very promising direction for future developments and might finally lead to robotic devices that "feel good".

References

1. Giummarra, M.J., Gibson, S.J., Georgiou-Karistianis, N., Bradshaw, J.L.: Mechanisms underlying embodiment, disembodiment and loss of embodiment. Neurosci. Biobehav. Rev. **32**, 143–160 (2008)
2. Pazzaglia, M., Molinari, M.: The embodiment of assistive devices—from wheelchair to exoskeleton. Phys. Life Rev. **16**, 163–175 (2016)
3. Botvinick, M., Cohen, J.: Rubber hands 'feel' touch that eyes see. Nature **391**, 756 (1998)
4. Beckerle, P., Salvietti, G., Unal, R., Prattichizzo, D., Rossi, S., Castellini, C., Hirche, S., Endo, S., Ben Amor, H., Ciocarlie, M., Mastrogiovanni, F., Argall, B.D.: A human-robot interaction perspective on assistive and rehabilitation robotics. Front. Neurorobotics **11**, 24 (2017)
5. Beckerle, P., Castellini, C., Lenggenhager, B.: Robotic interfaces for cognitive psychology and embodiment research: a research roadmap. Wiley Interdiscip. Rev.: Cogn. Sci. **10**(2), e1486 (2019)
6. Toet, A., Kuling, I.A., Krom, B.N., van Erp, J.B.F.: Toward enhanced teleoperation through embodiment. Front. Robot. AI **7**, 14 (2020)
7. Christ, O., Reiner, M.: Perspectives and possible applications of the rubber hand and virtual hand illusion in non-invasive rehabilitation: technological improvements and their consequences. Neurosci. Biobehav. Rev. **44**, 33–44 (2014)
8. Makin, T.R., de Vignemont, F., Faisal, A.A.: Neurocognitive barriers to the embodiment of technology. Nat. Biomed. Eng. **1**(1), 1–3 (2017)
9. Rognini, G., Blanke, O.: Cognetics: robotic interfaces for the conscious mind. Trends Cogn. Sci. **20**(3), 162–164 (2016)
10. Beckerle, P., De Beir, A., Schürmann, T., Caspar, E.A.: Human body schema exploration: analyzing design requirements of robotic hand and leg illusions. In: IEEE International Symposium on Robot and Human Interactive Communication (2016)
11. Hatzfeld, C., Kern, T.A.: Engineering Haptic Devices. Springer (2016)

Chapter 2
Concepts, Potentials, and Requirements

Abstract Understanding the embodiment of robotic devices and using this knowledge to improve human-robot interaction touches a variety of open research questions. A considerable body of research outlines the complexity and plasticity of human bodily experience. When examining robotic devices, which can be seen as "intelligent" tools, this challenge is getting even tougher due to the interaction of the involved agents. Beyond this, recent studies point out how the investigation of psychological fundamentals can benefit from using robotic devices for human-in-the-loop evaluation in return. This chapter conveys and discusses fundamental concepts and terminology of human body experience, aiming at accessibility for all concerned disciplines. Concepts of different body representations and the presence in virtual environments are presented along with fundamentals of human motor control and haptic perception. Based on those, perspective potentials in human-robot interaction are outlined with respect to control, sensing, feedback, and assessment. To support the design and application of human-in-the-loop approaches in fundamental research and engineering, the related literature is analyzed to determine and assess crucial design requirements.

2.1 Body Representations, Presence, and Their Intersections

This section presents working definitions of psychological terms and concepts regarding human body experience. A key aspect is body representations that try to grasp how humans experience their bodies. Psychological and neuroscientific literature distinguish various concepts of body representations, which are still subject to active research and controversial discussion. For simplicity, this monograph differentiates the concepts shown in Fig. 2.1: "body schema" and "body image" [1, 2]. Moreover, presence as the "feeling of being there" when operating in virtual environments is considered since virtual reality techniques are promising tools to develop future experiments. Finally, the intersections of the concepts are discussed based on a recent systematic literature review [3].

P. Beckerle, *Human-Robot Body Experience*, Springer Series on Touch and Haptic Systems, https://doi.org/10.1007/978-3-030-38688-7_2

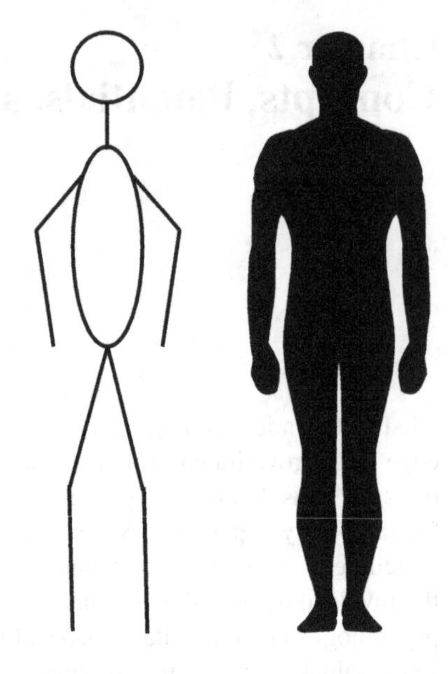

Fig. 2.1 Conceptual sketches of fundamental body representations: body schema (left) and body image (right)

2.1.1 Body Schema

The body schema concept describes a subconscious, neurophysiological, and multisensory representation of the characteristics of one's own body [2, 4]. It might be understood as an internal model of the body mechanics as indicated by the sketch in Fig. 2.1 and can integrate passive tools, but also robotic limbs [5–7]. As kinematic and dynamic properties are represented by the body schema, it is linked to the sense of having control over the own body, which makes it a potential benchmark for human-robot interface design [8]. Yet, psychological consequences of amputations indicate that the body schema is also related to well-being and quality of life and thus goes far beyond limb functionality [9]. Apparently, only well-designed technical aids are embodied and accepted by users [10, 11].

In 1998, a seminal paper by Botvinick and Cohen [12] initiated a line of research that is of the highest interest regarding psychological and neuroscientific fundamentals of the human mind and bodily self-experience. It reports how a feeling of body ownership towards a rubber hand, which is visible to the participant, can be elicited if this fake hand and the participant's hidden real hand are synchronously stimulated by brush strokes (see Fig. 2.2) [12]. This embodiment effect, i.e., the rubber hand illusion (RHI), seems to be caused by multisensory integration of vision, touch, and proprioception as well as their temporal and spatial stimulation [2, 8, 11, 13]. Many studies evaluate the body schema integration of the fake hand by objective measures, e.g., the proprioceptive drift as the perceived position of the real hand before and after stimulation [14, 15], and subjective measures, e.g., the well-established questionnaire

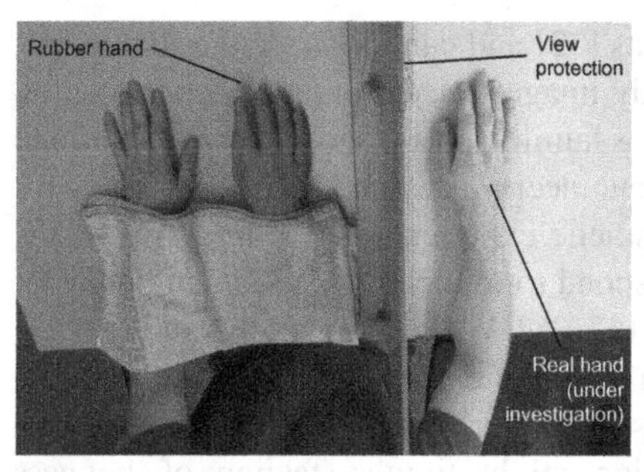

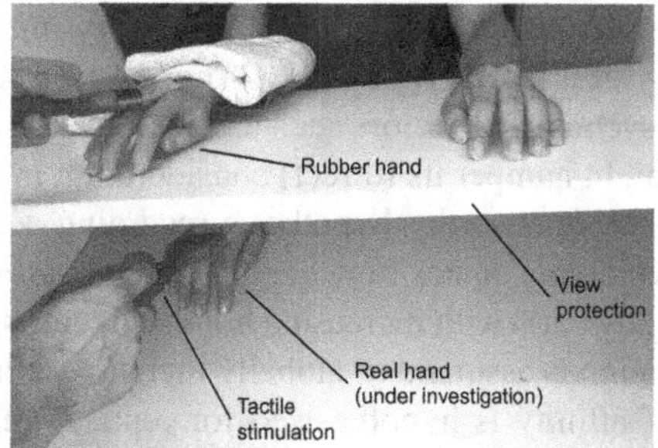

Fig. 2.2 Examples of rubber hand illusion experiments with horizontal (left) and vertical (right) alignment of the real hand that is under investigation and the fake hand. Tactile stimulation is exemplarily shown in the vertical setup (right). *Based on pictures provided by Dr. sc. hum. Robin Bekrater-Bodmann*

by Longo et al. [16]. This questionnaire analyzes the experience of embodiment considering the subcomponents agency, location, and ownership: agency describes the feeling to be in control of the fake limb, location measures its spatial experience and ownership rates the perceived belongingness to the participant. Meanwhile, a large body of research examines, which modulating factors (=modulators) influence the illusionary effect, and transferred bodily illusions to other limbs [17–20] and the whole body [21–23]. However, the underlying mechanisms are still not completely understood, e.g., because the perceptual and neuronal stability of the illusions might also depend on the experimental setup [24].

Recently, robotic limbs and virtual environments showed to be very promising approaches to examine various kinds of bodily illusions in technically augmented experiments [8, 11, 25]. Such human-in-the-loop paradigms are referred to as robotic hand/leg illusions (RobHI/RobLI) [8, 26–28] or virtual hand/leg illusions [11, 29]. They open up new experimental possibilities [25]: factors that influence body experience can be modified very selectively, e.g., spatial and temporal variations of visuotactile stimuli [11], stimulation patterns can deviate from natural ones, e.g., random finger reactions [27, 30], and different feedback paths can be altered individually, e.g., separating visual from tactile feedback [25, 31]. Especially, "unnatural" stimulation possibilities are very promising since they have the potential to improve the understanding of the psychological and neuroscientific fundamentals by disentangling action-perception loops, which cannot be changed in classical experimental paradigms [25].

2.1.2 Body Image

The body image is a representation of how humans experience their own outer appearance visually (see Fig. 2.1), but also psychologically, i.e., including their emotions and attitudes [2, 32]. As for the body schema, this first-person experience is based on the processing of multiple sensory inputs [32] and can influence the functionality of artificial limbs [33, 34]. Yet, the body image concept is further connected to psychosocial factors such as the influence of familiy, peers, and society [35], which might hamper its (direct) consideration in engineering design [34, 36].

A noteworthy hypothesis regarding aesthetic experience is the uncanny valley, which postulates how human observers respond to replications of human bodies or body parts with increasing human-likeness [37, 38]. While affinity towards the replication is assumed to globally increase with rising human-likeness, a local minimum of affinity is hypothesized for replications, which are only close to a human-like appearance. Those nearly human-like solutions might stimulate feelings of alienness and revulsion. Since the initial paper of Mori, various experimental studies observed uncanny valley-like effects practically [39], but its occurrence and its exact progression are still disputable [40, 41]. Despite the remaining unclarities, the related studies outline how aesthetic aspects of robot appearance might be quantified to be systematically considered in the design of robotic devices [39–41].

2.1.3 *Presence*

Presence conceptually describes the subjective experience of how users perceive being in a virtual environment [42–44] and can briefly be defined as the sense of "being there" [45]. While definitions are ambiguous [46] and psychometric measures are discussed [3], the concept of presence is technically promising since it relates to perceived interaction capabilities of users in virtual environments [42–44]. A very important example of such mediated actions is teleoperation [47–50]. While teleoperation systems can be assessed considering quantitative criteria of human-in-the-loop task performance, e.g., task-completion time or accuracy [48, 51], their design might also benefit from taking into account subjective presence experience [3].

This monograph adopts the working definition by Slater [52] that assumes someone to feel present when "perceptual, vestibular, proprioceptive, and autonomic nervous systems are activated in a way similar to that of real life in similar situations". Besides explaining the concept of presence, this definition emphasizes the rather challenging technical requirements. Similar to embodiment, presence exhibits various subcomponents of which self-presence, i.e., the extension of the self through virtual self-representations [53], relates presence to bodily self-experience and thus to the concepts of body schema or body image [54].

2.1.4 *Intersections*

To analyze connections between the concepts of embodiment and presence, a systematic literature review focusing on mediated scenarios and, especially, teleoperation was conducted by Nostadt et al. [3]. Interestingly, only 14 out of 414 related articles satisfied the purposely strict exclusion and inclusion criteria asking whether relations of embodiment and presence were discussed theoretically or examined experimentally considering appropriate measures.

The two identified theoretical frameworks relate presence to body representations in general: Haans and Ijsselsteijn [55] assume that a dynamic body schema is necessary to experience presence and mention that body image might have an influence. Kilteni et al. [56] consider that embodiment and self-presence might be used synonymously and that the presence might relate to (self-)location. This substantiates the hypothesized relations between the embodiment and presence concepts but also points out that their specific intersections are not fully understood [3]. Remarkably, the 12 reviewed experimental studies do not suffice to confirm or reject the validity of the theoretical frameworks due to definitional problems and non-standardized assessment methods [3].

Despite these intersections, embodiment theory seems to be a promising approach to assess highly immersive, anthropomorphic teleoperation in particular [3]. As suggested by Nostadt et al. [3], creating a comprehensive framework of presence and body representations as well as teleoperation system design might benefit from three guiding principles: mechanical fidelity, spatial bodily awareness, and self-identification.

Mechanical fidelity aggregates system properties, e.g., performance, accuracy, or dexterity, and is hence associated with the sense of agency [56] and with body schema integration [55]. While providing mechanically valid interaction represents a fundamental objective of teleoperation system design, spatial bodily awareness, and self-identification relate to higher aspects of body schema and body image [55, 56] in terms of embodied cognition [57] and might also influence task execution [3, 55].

2.2 Human-Robot Interaction Potentials

Looking at applications, the increasingly tight interaction of humans and robots makes the human-robot body experience a very prevalent topic. Beyond human-robot collaboration, wearable robots, e.g., for assistance and rehabilitation, are required to support and augment their users [58]. This addresses recent societal needs in healthcare [59, 60] as well as industrial developments [59, 61] and substantiates the importance of understanding and considering human-robot body experience [58, 62].

Remarkably, certain limitations in the effectiveness of assistive robotics are not caused by technical constraints, but by insufficient knowledge about human-robot interaction [59–61]. Hence, the development of such robotic devices appears to require human-oriented approaches, which consider technical and human aspects by combining methods from human and technical sciences [58].

This section points out existing challenges in human-robot interaction, especially concerning wearable robotics, and relates them to the concept of human-robot body experience based on [58]. It also puts those challenges into broader context considering the knowledge of human motor control and haptic perception.

2.2.1 Motor Control and Machine Learning

Ideally, wearable robots should "fit like a glove" but also be very easy to use [58]. This directly relates human-robot body experience to human motor control, which influences human-robot interaction and is an important determinant of ease-of-use. Since human motor control relies on complex coordination of processes in brain, muscles, and limbs interacting with the environment [63], robotic interaction behavior and assistance need to be tailored to the target application scenario user population [58, 63]. To this end, engineering research has to look beyond its horizon and consider psychological, physiological, and computational approaches [63]. Overall, aligning interaction strategies with human motor control is challenging since core problems multiply: human and robot agents have to deal with controlling many degrees of freedom, needing to sequence and time actions and integrate action-perception loops [63]. Moreover, they need to coordinate their intentions and actions to reach common aims and might both learn at the same time [58, 63].

Remarkably, many approaches tackling these challenges can similarly be applied in robotics as they appear to occur in humans. For instance, so-called kinematic synergies, which map (human) motions and actions with many degrees of freedom to a low-dimensional representation, can be mimicked and exploited to simplify robot motion control [64–66]. Since its conception, the concept of synergies in human motor control has even been extended to force, muscle, and neural levels [65]. Further, integrating actions and perception relies on sensory feedback to create closed-loop interactions and to mitigate errors in movement execution and coordination [25, 63]. Therefore, motor control can use dynamics models to process a copy of the motor command, i.e., an efference copy, to anticipate and shape a movement [67]. In the control of arm movements, this reafference principle was shown to cancel out effects of self-motion on sensation and thereby support discriminating whether a motion is induced by oneself or externally [67, 68]. Another very important example of perceptual-motor integration is mirror neurons, which show similar neural activity patterns for observing an action and for performing it oneself [69]. This mirror system might constrain body representations to support motion imitation and learning [70]. To enable motion learning, the human brain exhibits neural plasticity that enables it to adapt to changing requirements and environments based on personal experience [63]. This appears to link to the reference principle, which apparently relies on learned models of the own body [67].

Aiming to improve (tight) human-robot interaction, many control approaches have been suggested to reduce cognitive load and training efforts, among which semi-autonomous methods and machine learning recently show promising potential [58]. Shared control allows to offload cognitive and physical load due to semi-autonomous operation of the assistive device [71, 72] or teleoperation system [48]. Yet, the question of how to appropriately distribute control to the robotic system and the human is not finally answered: autonomous behavior needs to be predictable for users, provide appropriate assistance at the right time, and should not override users' intentions [58, 73]. Thus, intention detection and shared control algorithms should be designed considering human-robot body experience, which might be supported by means of machine learning [25, 58, 74, 75]. Learning approaches can enable user-specific adjustment of the robotic devices and lead to incremental co-adaptation between user and device exploiting neural plasticity and behavioral adaptation of the human [58, 76].

2.2.2 Haptic Perception and Sensory Feedback

Beyond the general value of sensory feedback for action-perception integration in human motor control [63], various studies suggest using force and tactile feedback in semi-autonomous [77–79] as well as in learning control [75] to convey information through physical interaction.

Taking a fundamental look shows that the human sense of touch relies on a multitude of peripheral sensory receptors [80]. Those comprise mechanoreceptors, which capture mechanical information in the skin, i.e., cutaneous receptors for tactile information, as well as in muscles, tendons, and joints, i.e., kinesthetic receptors [80, 81]. Furthermore, thermoreceptors provide information about the thermal conditions at the skin surface [80]. The remainder of this book focuses on the mechanical aspect that is categorized in active and passive interactions: while active haptics describes sensations and interactions due to conscious and controlled user movements with low frequencies, the major part of haptic perception is passive [81]. This part covers a wide range of frequencies at which users do not respond actively, but rather (slowly) adapt their mechanical interaction properties [81].

The particular sensations and responses depend on the mechanical body characteristics of an individual, which are described by linear, time-invariant mechanical transfer functions of how the system resits motion, i.e., impedances [81]. Besides being user-specific, impedance characteristics depend on actions, e.g., current grasp types [81] and in which region of the body the haptic interaction occurs [80, 82]. For cutaneous receptors, one should distinguish those embedded in hairy and hairless skin, i.e., glabrous skin, of which the latter is focused by research up to now [80].

Shaping a user's experience, cutaneous and kinesthetic sensations are weighted and fused, also crossmodally with other percepts, to finally be processed on a cognitive level [80, 83, 84]. Through multisensory information integration and cognitive processing, humans are capable of discriminating objects, surfaces, and their properties [80]. Moreover, haptic experiences serve identifying stimulus locations and localizing the own body in relation to physical contact partners, i.e., spatial perception [80], which directly links it to human body experience and body representations. Beyond these rather functional aspects of haptics, affective aspects such as pleasantness and emotional information receive increased research interest [80, 85] and its receptors are, interestingly, mainly located in hairy skin [86]. Moreover, it has to be noted that the sense of touch, similarly to motor control, is subject to neural plasticity and can therefore adapt to changing interaction scenarios and environments [80].

To cover all aspects of providing high-fidelity haptic feedback, a lot of promising technologies was developed recently, e.g., large-surface sensorized robotic skins [87–89] and wearable haptic stimulators [90, 91]. Despite having the potential to close the action-perception loop of user and robot, these new technologies did not yet find common application in wearable/assistive robotics [58]. Interestingly, the example of prosthetic devices shows that such lacking afferent sensory information can be a major cause of device abandonment [92–94].

As analyzed by Beckerle et al. [58], rendering realistic haptic feedback is technically not yet reached and, furthermore, suffers from missing understanding of human experience. Recent studies mention first positive results with high-fidelity sensory feedback [95], but issues like information overload and performance degradation are observed as well [58, 96]. Still, psychological research agrees that appropriate feedback is the basis for successful multisensory integration and facilitation of device embodiment [10, 97]. A possible strategy to provide natural and consistent haptic feedback with high mechanical fidelity [3] could be sensory synergies that map low-level sensory variables to high-level cognitive percepts [98]. Based on knowing humans implement this mapping, we might be able to set up mathematical models of haptic perception and implement them in intuitive interfaces [58, 99, 100]. Additionally, future human-machine interfaces might consider effective information and their influence on social and emotional interaction [85]. Beyond established approaches from psychology, first approaches to specifically evaluate and measure emotional aspects of haptic stimuli in interface design exist [101, 102].

2.2.3 Assessment Metrics and Methods

Beyond the progress in control and sensory feedback, human-centric standards to evaluate tightly interacting robotic devices appear to be missing [58]. Taking a look at healthcare scenarios, assessing user improvements that are due to the robot is challenging [103] and tools for quantitative functional assessment in a wide range of activities are missing [58]. Despite the complexity due to the variety of users, activities, and devices [59–61], future assessment procedures could adapt classical tests continuously during interaction and thereby tackle the upcoming challenges due to incremental learning and shared autonomy techniques [58, 76]. In this respect, evaluating human-robot body experience is a very promising metric to improve the wearability and intuitivity of robotic devices and their interfaces [25, 58]. To comply with the individual abilities and preferences of each user, interaction strategies need to be tailored to the joint task [58]. Human-in-the-loop experiments are a promising approach to individualize developments early on during human-centered design approaches [34] and in benchmarking existing systems [104].

Fig. 2.3 Experimental robotic hand illusion setup (extended version used in [31, 115] and based on the design from [27, 107]). *Based on a figure from* Huynh et al. [31]

2.3 Human-in-the-loop System Design Requirements

Human-in-the-loop experiments enable human participants to experience mechatronic and robotic technology [25, 73, 105]: participants are equipped with either the investigated device or a technical simulator to become part of the human-machine system and its control loop. Since the devices/simulators themselves are currently facing technical limitations, e.g., intrinsic time delays [106, 107], compromises need to be made during system design. This demands a precise analysis of system requirements considering the influence of the human user and, in return, might provide helpful information for the development of the target application. A technical requirement analysis of human-in-the-loop systems designed to explore human-robot body experience, which was conducted in the interdisciplinary expert discussion by Beckerle et al. [8], is presented subsequently. By juxtaposing upper limb and one lower limb systems, joint design requirements and suggestions for system development are derived.

2.3.1 Robotic Approaches

To go beyond "active" versions of the rubber hand illusion [108, 109], which mechanically transfer human movements to the fake hand, different research groups employed robotic limbs to investigate bodily illusions [27, 28, 31, 110–113]. Being programmable, robotic limbs facilitate to re-configurably and reproducibly disentangle human action and perception loops by manipulating robot control and sensory feedback, e.g., control delays or haptic rendering [8, 25]. While virtual reality methods might allow similar or even more variability, the complexity of designing robotic human-in-the-loop systems also helps to improve technical implementation [25, 114].

The *Robotic Hand Illusion* (RobHI, see Chap. 3) replaces the rubber hand with a robotic hand that is controlled through a human-machine interface. Figure 2.3 shows the setup with the hidden real hand wearing the human-machine interface (left) and the robotic hand (middle). The specific setup presented in Fig. 2.3 measures the users' finger motions using a sensory glove with flex sensors and commands servo motors to drive the 3D-printed hand/finger mechanisms correspondingly [107]. The design was later extended to provide vibrotactile feedback at the fingertips [115]. In contrast to commercially available robotic hands that are optimized to achieve certain grasping forces, the key design requirements were human-like shape and fast finger motions [8].

The *Robotic Leg Illusion* (RobLI, see Chap. 5) transfers the bodily illusion paradigm to the legs. Figure 2.4 shows a rear view of the setup with the instrumented hidden real leg (middle) and the robot leg (right). The particular implementation depicted in Fig. 2.4 acquires sagittal plane squatting motions of the user's hidden ankle and knee joint by inertial measurement units (IMUs) attached to an interface orthosis with integrated vibrotactile feedback stimulators [116, 117]. Feedback-controlled DC motors drive the knee and ankle joints of the robot leg that exhibits adjustable segment lengths [112].

2.3.2 Design Requirements

The requirements are clustered into setup characteristics, robot mechanics, robot actuation, robot control, human motion acquisition, and user feedback [8]. Setup characteristics requirements are presented in Table 2.1, while Tables 2.2 and 2.3 analyze those relating to the robot and the interface. The tables also outline, which RobHI and RobLI design requirements were judged to be similar (indicated by a "*" in the criterion column) or different, and thereby point out the consistent requirements of the different bodily illusions [8].

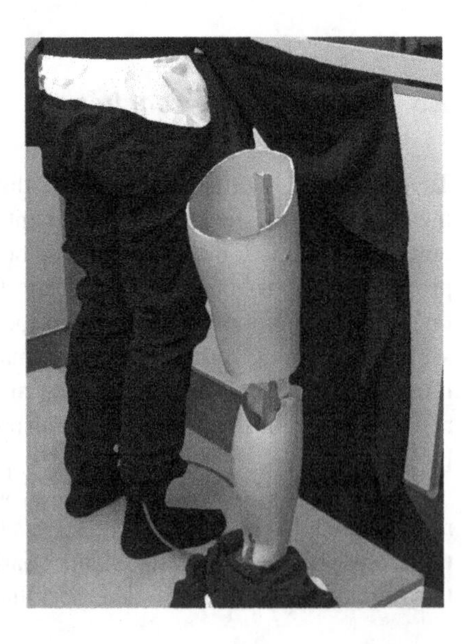

Fig. 2.4 Experimental robotic leg illusion setup [112, 113, 116]. *Based on a figure from* Beckerle et al. [8]

Table 2.1 Comparison of RobHI and RobLI setup requirements based on [8]

Cluster	Criterion	RobHI	RobLI
Setup characteristics	Distance of human and robot*	Max. 0.275 m [118], position of real and fake hand relatively to the body is important [119]	0.1–0.2 m [112, 118]
	Hiding the human limb	Hands separated by rigid structure, human arm covered by fabric	Legs covered by frame with fabric, challenging during squats [117]
	Anatomical plausibility*	Robotic hand position and orientation should be anatomically plausible [119, 120]	Robot can be adapted to population [112], orientation appears relevant
	Transportability*	Robotic hand setup should be transportable [107]	Transportable solution required [112]
	Visual appearance of robotic device*	Similar to human hand, but illusion is quite robust regarding small visual variations [121]	Shape, laterality, and cladding should be human-like [112, 122]
	Software-controlled experiments*	Automatic procedure to avoid experimenter bias and remove variability	Digital control to reduce variability and bias through the experimenter
	Reliability and robustness*	Relevant due to frequent use by (non-)experts	Should be high to focus on research

Setup characteristics represent general characteristics of the experiment and are described in Table 2.1. Key requirements are the distance between human and robot, anatomical plausibility, and hiding the real human limb [8]. For RobHI and RobLI, the distance of human and robotic hand/leg is similarly quantified with values below 0.3 m and anatomical plausibility demands for human-like limb orientation [117, 119]. Contrary, hiding the real limb might be more complex in case of the RobLI due to the need for weight bearing, e.g., if considering the squatting scenario [112]. From a global perspective, the robotic device has to fulfill requirements regarding its visual appearance, reliability and robustness, the option to transport it, and automizing the experimental procedure. While visual appearance should be human-like, software-controlled experiments and computer-generated instructions help avoiding human-caused biases.

Requirements regarding the engineering design of the robotic device itself are given in Table 2.2 and concern mechanics, actuation, and control. The mechanical requirements are similar for hands as well as legs and allow for certain deviations from ideal biomimicry, e.g., finger underactuation or limitation to sagittal plane [8]. Adjustability of the robot to the body geometries of particular participants might be more relevant with respect to the leg [119]. Paramountly, the motion dexterity required for the investigated task needs to be covered, which also links to the comparable actuation speed requirements of RobHI and RobLI. Due to weight bearing,

Table 2.2 Comparison of requirements regarding the robotic device based on [8]

Cluster	Criterion	RobHI	RobLI
Robot mechanics	Degrees of freedom*	Underactuated flexion for each finger is sufficient [107], thumb rarely used	Limitation to rotational ankle and knee joints in sagittal plane [112]
	Range of motion*	Small to complete flexion/extension for each finger [8]	90° for the ankle and 180° for the knee [112]
	Adjustability to human body geometry	Standard dimensions due to illusion robustness, size sensory glove might be adjusted	Variable adaptation to participants body geometries possible [112]
	Weight*	Light digits to reduce inertia and increase motion speed [107]	Lightweight to minimize actuation effort [112]
Robot actuation	Speed*	High speed for minimum delays of finger flexion up to 3 Hz	High speed with minimum delay up to 1 Hz [112, 124]
	Torque/force	Low torque and/or compliant actuation for user safety	Depends on the robot mass, rigid actuation for position tracking [112]
	Power*	Low voltage for safety and to reduce electromagnetic disturbances	Knee: low to reduce weight; ankle: high to reduce load for better acoustics [112]
	Size and weight*	No specific requirements when using tendons [107], small actuators if directly integrated [28]	Lightweight to reduce motor loads [112] or remote actuation [113]
	Acoustics*	Shielding and damping to reduce acoustic noise	Shielding and appropriate actuator load/dimensioning to avoid acoustic disturbance [112]
Robot control	Spatial accuracy	Low, but constant position shift acceptable [107]	Similar trajectories up to 1 Hz [112]
	Temporal accuracy*	Small, constant and quantifiable delay [107]	Below 0.1 s including motion acquisition [112, 125]
	Robot position sensing	Position sensing via servo motors	Required, standard encoders sufficient [112]
	Motion suppression at rest*	No trembling movements	No trembling movements, smooth motions required
	Motion (re-)mapping	Possibility to set precise delays and randomize motions [27]	Delays should be programmable [112], no re-mapping

the force/torque requirements of the (squatting) RobLI are higher, but human safety should be of the highest priority in both cases. To achieve a natural appearance and avoid interferences, low-power, and lightweight actuators are required and acoustic emission should be considered [123].

System accuracy and delay strongly depend on robot as well as interface requirements. The latter are presented in Table 2.3 and include human motion acquisition

Table 2.3 Comparison of RobHI and RobLI interface design requirements based on [8]

Cluster	Criterion	RobHI	RobLI
Human motion acquisition	Spatial accuracy	Position of real and fake fingers can differ [107, 120]	Congruent motions [106, 112]
	Temporal accuracy*	Small, constant an precisely measurable delays	Not higher than 0.1 s combined with those due to motion control [106, 112, 125]
	Measurement noise*	Less an issue using flex sensor	Rather uncritical considering IMUs [117]
	Size and weight (of worn components)*	Devices should not interfere with the hand motion	Should not disturb motions [112], but contact to the body is acceptable [116, 117]
Sensory feedb	Perceptual channels*	Vibrotactile and acoustic feedback might be considered	Vibrotactile through vibration motors to investigate the influence of tactile feedback on multisensory integration [116]

and sensory feedback. Low system-intrinsic delays are a major requirement to induce all bodily illusions and put tough constraints on control and motion acquisition algorithms [8]. Moreover, quantifiable delay extensions are necessary for most experimental paradigms since asynchronous conditions are required to show whether an illusion occurs in the synchronous ones. Contrary, spatial accuracy appears less important than reaction speed, but might be more relevant in RobLI due to the wider range of motions.

Among the interface requirements in Table 2.3, instrumentation for human motion acquisition should neither cause technical nor experiential disturbances. For instance, robot (jittering) motions should be suppressed when the participant's limb rests to avoid disturbances of the illusion [107]. Sensors should provide low-noise output and unobtrusive in terms of size and weight. Since tactile stimulation can have a considerable contribution to the illusionary effect, haptic feedback is interesting in both RobHI and RobLI [8].

2.3.3 Design Implications

The five most crucial design factors identified in the analysis of Beckerle et al. [8] are hiding the real limb, anatomical plausibility, visual appearance, temporal delay, and software-controlled experimental conditions. Since the system-intrinsic delay due to robot control and human motion acquisition represents a general system property, all top five requirements can be related to setup characteristics that determine whether illusions can be elicited. Similarly, other high-ranked criteria influence the quality of the underlying psychological effect, e.g., acoustics or the suppression of robotic motion during human rest. For experiments, a constant and precisely quantifiable temporal delay appears to be of high importance. Concluding, experience-related setup requirements should be prioritized in human-in-the-loop systems engineering.

Interpreting these design implications from a psychological perspective implies a strong relation to the crossmodal processing and integration of vision, touch, and proprioception, which underlies bodily illusions. Interestingly, the experimental elicitation of bodily illusions is possible in fractions of minutes [126], while patients with amputation need months to develop a perception of embodiment over the prosthesis [127], which might indicate the need for further in-the-field experimentation.

References

1. Kammers, M.P.M., van der Ham, Ineke J.M, Dijkerman, H.C.: Dissociating body representations in healthy individuals: differential effects of a kinaesthetic illusion on perception and action. Neuropsychologia **44**(12), 2430–2436 (2006). https://doi.org/10.1016/j.neuropsychologia.2006.04.009
2. Mayer, A., Kudar, K., Bretz, K., Tihanyi, J.: Body schema and body awareness of amputees. Prosthet. Orthot. Int. **32**(3), 363–382 (2008)
3. Nostadt, N., Abbink, D.A., Christ, O., Beckerle, P.: Embodiment, presence, and their intersections: teleoperation and beyond (submitted). ACM Trans. Hum. Robot Interact. (2020)
4. Gallagher, S., Cole, J.: Body schema and body image in a deafferented subject. J. Mind Behav. **16**, 369–390 (1995)
5. Holmes, N.P., Spence, C.: The body schema and the multisensory representation(s) of peripersonal space. Cogn. Process. **5**(2), 94–105 (2004)
6. Maravita, A., Iriki, A.: Tools for the body (schema). Trends Cogn. Sci. **8**(2), 79–86 (2004)
7. Cardinali, L., Frassinetti, F., Brozzoli, C., Urquizar, C., Roy, A.C., Farnè, A.: Tool-use induces morphological updating of the body schema. Curr. Biol. **19**(12), R478–R479 (2009)
8. Beckerle, P., De Beir, A., Schürmann, T., Caspar, E.A.: Human body schema exploration: analyzing design requirements of robotic hand and leg illusions. In: IEEE International Symposium on Robot and Human Interactive Communication (2016)
9. Senra, H., Aragao Oliveira, R., Leal, I., Vieira, C.: Beyond the body image: a qualitative study on how adults experience lower limb amputation. Clin. Rehabil. **26**, 180–191 (2011)
10. Giummarra, M.J., Gibson, S.J., Georgiou-Karistianis, N., Bradshaw, J.L.: Mechanisms underlying embodiment, disembodiment and loss of embodiment. Neurosci. Biobehav. Rev. **32**, 143–160 (2008)
11. Christ, O., Reiner, M.: Perspectives and possible applications of the rubber hand and virtual hand illusion in non-invasive rehabilitation: technological improvements and their consequences. Neurosci. Biobehav. Rev. **44**, 33–44 (2014)
12. Botvinick, M., Cohen, J.: Rubber hands 'feel' touch that eyes see. Nature **391**, 756 (1998)
13. Haggard, P.: Conscious intention and motor cognition. Trends Cogn. Sci. **9**(6), 290–295 (2005)
14. Tsakiris, M., Haggard, P.: The rubber hand illusion revisited: visuotactile integration and self-attribution. J. Exp. Psychol.: Hum. Percept. Perform. **31**(1), 80–91 (2005)
15. Makin, T.R., Holmes, N.P., Ehrsson, H.H.: On the other hand: dummy hands and peripersonal space. Behav. Brain Res. **191**(1), 1–10 (2008)
16. Longo, M.R., Schüür, F., Kammers, M.P.M., Tsakiris, M., Haggard, P.: What is embodiment? A psychometric approach. Cognition **107**, 978–998 (2008)

17. Christ, O., Elger, A., Schneider, K., Beckerle, P., Vogt, J., Rinderknecht, S.: Identification of haptic paths with different resolution and their effect on body scheme illusion in lower limbs. Technically Assisted Rehabilitation (2013)
18. Lenggenhager, B., Hilti, L., Brugger, P.: Disturbed body integrity and the "rubber foot illusion". Neuropsychology **29**(2), 205 (2015)
19. Crea, S., D'Alonzo, M., Vitiello, N., Cipriani, C.: The rubber foot illusion. J. Neuro Eng. Rehabil. **12**, 77 (2015)
20. Flögel, M., Kalveram, K.T., Christ, O., Vogt, J.: Application of the rubber hand illusion paradigm: comparison between upper and lower limbs. Psychol. Res. **80**(2), 298–306 (2015)
21. Lenggenhager, B., Tadi, T., Metzinger, T., Blanke, O.: Video ergo sum: manipulating bodily self-consciousness. Science **317**(5841), 1096–1099 (2007)
22. Aspell, J.E., Lenggenhager, B., Blanke, O.: Keeping in touch with one self: multisensory mechanisms of self-consciousness. PLoS ONE **4**(8), (2009)
23. Maselli, A., Slater, M.: The building blocks of the full body ownership illusion. Front. Hum. Neurosci. **7**, 83 (2013)
24. Bekrater-Bodmann, R., Foell, J., Diers, M., Flor, H.: The perceptual and neuronal stability of the rubber hand illusion across contexts and over time. Brain Res. **1452**, 130–139 (2012)
25. Beckerle, P., Castellini, C., Lenggenhager, B.: Robotic interfaces for cognitive psychology and embodiment research: a research roadmap. Wiley Interdisc. Rev.: Cogn. Sci. **10**(2), e1486 (2019)
26. Ehrsson, H.H., Rosén, B., Stockselius, A., Ragnö, C., Köhler, P., Lundborg, G.: Upper limb amputees can be induced to experience a rubber hand as their own. Brain **131**(12), 3443–3452 (2008)
27. Caspar, E.A., de Beir, A., Magalhães Da Saldanha da Gama, P.A., Yernaux, F., Cleeremans, A., Vanderborght, B.: New frontiers in the rubber hand experiment: when a robotic hand becomes one's own. Behav. Res. Methods **47**(3), 744–755 (2015)
28. Romano, R., Caffa, E., Hernandez-Arieta, A., Brugger, P., Maravita, A.: The robot hand illusion: inducing proprioceptive drift through visuo-motor congruency. Neuropsychologia **70**, 414–420 (2015)
29. Pozeg, P., Galli, G., Blanke, O.: Those are your legs: the effect of visuo-spatial viewpoint on visuo-tactile integration and body ownership. Front. Psychol. **6** (2015)
30. Caspar, E.A., Cleeremans, A., Haggard, P.: The relationship between human agency and embodiment. Conscious. Cogn. **33**, 226–236 (2015)
31. Huynh, T.V., Bekrater-Bodmann, R., Fröhner, J., Vogt, J., Beckerle, P.: Robotic hand illusion with tactile feedback: unravelling the relative contribution of visuotactile and visuomotor input to the representation of body parts in space. PloS ONE **14**(1), e0210,058 (2019)
32. Piryankova, I.V., Stefanucci, J.K., Romero, J., De La Rosa, S., Black, M.J., Mohler, B.J.: Can i recognize my body's weight? the influence of shape and texture on the perception of self. ACM Trans. Appl. Percept. **11**(3) (2014)
33. Christ, O., Jokisch, M., Preller, J., Beckerle, P., Rinderknecht, S., Wojtusch, J., von Stryk, O., Vogt, J.: User-centered prosthetic development: comprehension of amputees' needs. Biomed. Eng. **57**(S1), 1098–1101 (2012)
34. Beckerle, P., Christ, O., Schürmann, T., Vogt, J., von Stryk, O., Rinderknecht, S.: A human-machine-centered design method for (powered) lower limb prosthetics. Robot. Auton. Syst. **95**, 1–12 (2017)
35. Breakey, J.W.: Body image: the inner mirror. JPO: J. Prosthet. Orthot. **9**(3), 107–112 (1997)
36. Gauthier-Gagnon, C., Grisé, M.C., D, P.: Enabling factors related to prosthetic use by people with transtibial and transfemoral amputation. Arch. Phys. Med. Rehabil. **80**(6), 706–713 (1999)
37. Mori, M.: The uncanny valley. Energy **7**(4), 33–35 (1970)
38. Mori, M., MacDorman, K.F., Kageki, N.: The uncanny valley [from the field]. IEEE Robot. Autom. Mag. **19**(2), 98–100 (2012)
39. Poliakoff, E., Beach, N., Best, R., Howard, T., Gowen, E.: Can looking at a hand make your skin crawl? peering into the uncanny valley for hands. Perception **42**(9), 998–1000 (2013)

40. Bartneck, C., Kanda, T., Ishiguro, H., Hagita, N.: Is the uncanny valley an uncanny cliff? In: IEEE International Symposium on Robot and Human interactive Communication, pp. 368–373. IEEE (2007)
41. Rosenthal-von der Pütten, A.M., Krämer, N.C.: How design characteristics of robots determine evaluation and uncanny valley related responses. Comput. Hum. Behav. **36**, 422–439 (2014)
42. Hendrix, C., Barfield, W.: Presence within virtual environments as a function of visual display parameters. Presence: Teleoperators Virtual Environ. **5**(3), 274–289 (1996). https://doi.org/10.1162/pres.1996.5.3.274
43. Sallnäs, E.L., Rassmus-Gröhn, K., Sjöström, C.: Supporting presence in collaborative environments by haptic force feedback. ACM Trans. Comput. Hum. Interact. **7**(4), 461–476 (2000). https://doi.org/10.1145/365058.365086
44. Lee, S., Kim, G.J.: Effects of haptic feedback, stereoscopy, and image resolution on performance and presence in remote navigation. Int. J. Hum. Comput. Stud. **66**(10), 701–717 (2008). https://doi.org/10.1016/j.ijhcs.2008.05.001
45. Heeter, C.: Being there: the subjective experience of presence. Presence: Teleoperators Virtual Environ. **1**(2), 262–271 (1992). https://doi.org/10.1162/pres.1992.1.2.262
46. Lombard, M., Jones, M.T.: Defining presence. In: Lombard, M., Biocca, F., Freeman, J., IJsselsteijn, W., Schaevitz, R.J. (eds.) Immersed in Media, pp. 13–34. Springer International Publishing, Cham (2015). https://doi.org/10.1007/978-3-319-10190-3.2
47. Lawrence, D.A.: Stability and transparency in bilateral teleoperation. IEEE Trans. Robot. Autom. **9**(5), 624–637 (1993). https://doi.org/10.1109/70.258054
48. Boessenkool, H., Abbink, D.A., Heemskerk, C.J.M., van der Helm, F.C.T., Wildenbeest, J.G.W.: A task-specific analysis of the benefit of haptic shared control during telemanipulation. IEEE Trans. Haptics **6**(1), 2–12 (2013)
49. Wildenbeest, J.G.W., Abbink, D.A., Heemskerk, C.J.M., van der Helm, F.C.T, Boessenkool, H.: The impact of haptic feedback quality on the performance of teleoperated assembly tasks. IEEE Trans. Haptics **6**(2), 242–252 (2013). https://doi.org/10.1109/TOH.2012.19
50. Pacchierotti, C., Tirmizi, A., Prattichizzo, D.: Improving transparency in teleoperation by means of cutaneous tactile force feedback. ACM Trans. Appl. Percept. **11**(1), 1–16 (2014). https://doi.org/10.1145/2604969
51. Pacchierotti, C., Meli, L., Chinello, F., Malvezzi, M., Prattichizzo, D.: Cutaneous haptic feedback to ensure the stability of robotic teleoperation systems. Int. J. Robot. Res. **34**(14), 1773–1787 (2015). https://doi.org/10.1177/0278364915603135
52. Slater, M.: A note on presence terminology. Presence Connect **3**(3), 1–5 (2003)
53. Ratan, R.A., Hasler, B.: Self-presence standardized: introducing the self-presence questionnaire (SPQ). In: 12th Annual International Workshop on Presence (2009)
54. Biocca, F.: The Cyborg's Dilemma: Progressive Embodiment in Virtual Environments, vol. 3. IEEE Computer Society Press, Los Alamitos, Calif. (1997). https://doi.org/10.1111/j.1083-6101.1997.tb00070.x
55. Haans, A., Ijsselsteijn, W.A.: Embodiment and telepresence: toward a comprehensive theoretical framework. Interact. Comput. **24**(4), 211–218 (2012). https://doi.org/10.1016/j.intcom.2012.04.010
56. Kilteni, K., Groten, R., Slater, M.: The sense of embodiment in virtual reality. Presence: Teleoperators Virtual Environ. **21**(4), 373–387 (2012). https://doi.org/10.1162/PRES.a.00124
57. Wilson, M.: Six views of embodied cognition. Psychon. Bull. Rev. **9**(4), 625–636 (2002)
58. Beckerle, P., Salvietti, G., Unal, R., Prattichizzo, D., Rossi, S., Castellini, C., Hirche, S., Endo, S., Ben Amor, H., Ciocarlie, M., Mastrogiovanni, F., Argall, B.D., Bianchi, M.: A human-robot interaction perspective on assistive and rehabilitation robotics. Front. Neurorobot. **11**(24) (2017)
59. Dollar, A.M., Herr, H.: Lower extremity exoskeletons and active orthoses: challenges and state-of-the-art. IEEE Trans. Robot. **24**(1), 144–158 (2008)
60. Windrich, M., Grimmer, M., Christ, O., Rinderknecht, S., Beckerle, P.: Active lower limb prosthetics: a systematic review of design issues and solutions. Biomed. Eng. Online **15**(3), 5–19 (2016)

61. Yan, T., Cempini, M., Oddo, C.M., Vitiello, N.: Review of assistive strategies in powered lower-limb orthoses and exoskeletons. Robot. Auton. Syst. **64**, 120–136 (2015)
62. Rognini, G., Blanke, O.: Cognetics: robotic interfaces for the conscious mind. Trends Cogn. Sci. **20**(3), 162–164 (2016)
63. Rosenbaum, D.A.: Human Motor Control. Academic press (2009)
64. Santello, M., Flanders, M., Soechting, J.F.: Postural hand synergies for tool use. J. Neurosci. **18**(23), 10105–10115 (1998)
65. Santello, M., Baud-Bovy, G., Jörntell, H.: Neural bases of hand synergies. Front. Comput. Neurosci. **7**, 23 (2013)
66. Santello, M., Bianchi, M., Gabiccini, M., Ricciardi, E., Salvietti, G., Prattichizzo, D., Ernst, M., Moscatelli, A., Jörntell, H., Kappers, A.M.L., Kyriakopoulos, K., Albu-Schäffer, A., Castellini, C., Bicchi, A.: Hand synergies: integration of robotics and neuroscience for understanding the control of biological and artificial hands. Phys. Life Rev. **17**, 1–23 (2016)
67. Wolpert, D.M.: Computational approaches to motor control. Trends Cogn. Sci. **1**(6), 209–216 (1997)
68. Iacoboni, M., Koski, L.M., Brass, M., Bekkering, H., Woods, R.P., Dubeau, M.C., Mazziotta, J.C., Rizzolatti, G.: Reafferent copies of imitated actions in the right superior temporal cortex. Proc. Nat. Acad. Sci. **98**(24), 13995–13999 (2001)
69. Rizzolatti, G., Craighero, L.: The mirror-neuron system. Annu. Rev. Neurosci. **27**, 169–192 (2004)
70. Pineda, J.A.: Sensorimotor cortex as a critical component of an 'extended' mirror neuron system: does it solve the development, correspondence, and control problems in mirroring? Behav. Brain Funct. **4**(1), 47 (2008)
71. Argall, B.D.: Modular and adaptive wheelchair automation. In: Proceedings of the International Symposium on Experimental Robotics (ISER) (2014)
72. Jain, S., Farshchiansadegh, A., Broad, A., Abdollahi, F., Mussa-Ivaldi, F.: Assistive robotic manipulation through shared autonomy and a body-machine interface. In: IEEE International Conference on Rehabilitation Robotics (2015)
73. Gopinath, D., Jain, S., Argall, B.D.: Human-in-the-loop optimization of shared autonomy in assistive robotics. IEEE Robot. Autom. Lett. **2**(1), 247–254 (2017). https://doi.org/10.1109/LRA.2016.2593928
74. Pilarski, P.M., Dawson, M.R., Degris, T., Carey, J.P., Chan, K.M., Hebert, J.S., Sutton, R.S.: Adaptive artificial limbs: a real-time approach to prediction and anticipation. IEEE Robot. Autom. Mag. **20**(1), 53–64 (2013)
75. Castellini, C., Artemiadis, P.K., Wininger, M., Ajoudani, A., Alimusaj, M., Bicchi, A., Caputo, B., Craelius, W., Došen, S., Englehart, K.B., Farina, D., Gijsberts, S., Godfrey, S.B., Hargrove, L.J., Ison, M., Kuiken, T.A., Markovic, M., Pilarski, P.M., Rupp, R., Scheme, E.: Proceedings of the first workshop on peripheral machine interfaces: going beyond traditional surface electromyography. Front. Neurorobot. **5**(22), 1–17 (2014)
76. Castellini, C., Bongers, R.M., Nowak, M., van der Sluis, C.K.: Upper-limb prosthetic myocontrol: two recommendations. Front. Neurosci. **9**(496) (2015). https://doi.org/10.3389/fnins.2015.00496
77. Dahiya, R.S., Metta, G., Valle, M., Sandini, G.: Tactile sensing - from humans to humanoids. IEEE Trans. Robot. **26**(1), 1–20 (2010)
78. Cannata, C., Denei, S., Mastrogiovanni, F.: A framework for representing interaction tasks based on tactile data. In: IEEE International Symposium on Robot and Human Interactive Communication (2010)
79. Mengüç, Y., Park, Y.L., Pei, H., Vogt, D., Aubin, P.M., Winchell, E., Fluke, L., Stirling, L., Wood, R.J., Walsh, C.J.: Wearable soft sensing suit for human gait measurement. Int. J. Robot. Res. **33**(14), 1748–1764 (2014)
80. Lederman, S.J., Klatzky, R.L.: Haptic perception: a tutorial. Atten. Percept. Psychophys. **71**(7), 1439–1459 (2009)
81. Hatzfeld, C., Kern, T.A.: Engineering Haptic Devices. Springer (2016)

82. Mancini, F., Bauleo, A., Cole, J., Lui, F., Porro, C.A., Haggard, P., Iannetti, G.D.: Whole-body mapping of spatial acuity for pain and touch. Ann. Neurol. **75**(6), 917–924 (2014)
83. Ernst, M.O., Banks, M.S.: Humans integrate visual and haptic information in a statistically optimal fashion. Nature **415**(6870), 429–433 (2002)
84. Helbig, H.B., Ernst, M.O.: Haptic perception in interaction with other senses. In: Human Haptic Perception: Basics and Applications, pp. 235–249. Springer (2008)
85. Beckerle, P., Kõiva, R., Kirchner, E.A., Bekrater-Bodmann, R., Dosen, S., Christ, O., Abbink, D.A., Castellini, C., Lenggenhager, B.: Feel-good robotics: requirements on touch for embodiment in assistive robotics. Front. Neurorobot. **12**, 84 (2018)
86. Crucianelli, L., Metcalf, N.K., Fotopoulou, A., Jenkinson, P.M.: Bodily pleasure matters: velocity of touch modulates body ownership during the rubber hand illusion. Front. Psychol. **4**, 703 (2013)
87. Muscari, L., Seminara, L., Mastrogiovanni, F., Valle, M., Capurro, M., Cannata, C.: Real-time reconstruction of contact shapes for large-area robot skin. In: IEEE International Conference on Robotics and Automation (2013)
88. Denei, S., Mastrogiovanni, F., Cannata, C.: Towards the creation of tactile maps for robots and their use in robot contact motion control. Robot. Auton. Syst. **63**, 293–308 (2015)
89. Youssefi, S., Denei, S., Mastrogiovanni, F., Cannata, C.: A real-time data acquisition and processing framework for large-scale robot skin. Robot. Auton. Syst. **68**, 86–103 (2015)
90. Prattichizzo, D., Chinello, F., Pacchierotti, C., Malvezzi, M.: Towards wearability in fingertip haptics: a 3-dof wearable device for cutaneous force feedback. IEEE Trans. Haptics **6**(4), 506–516 (2013)
91. Pacchierotti, C., Sinclair, S., Solazzi, M., Frisoli, A., Hayward, V., Prattichizzo, D.: Wearable haptic systems for the fingertip and the hand: taxonomy, review, and perspectives. IEEE Trans. Haptics **10**(4), 580–600 (2017)
92. Jiang, N., Došen, S., Müller, K., Farina, D.: Myoelectric control of artificial limbs - is there a need to change focus? IEEE Signal Process. Mag. **29**(5), 148–152 (2012)
93. Makin, T.R., de Vignemont, F., Faisal, A.A.: Neurocognitive barriers to the embodiment of technology. Nat. Biomed. Eng. **1**(1), 1–3 (2017)
94. Biddiss, E., Chau, T.: Upper-limb prosthetics: critical factors in device abandonment. Am. J. Phys. Med. Rehabil. **86**(12), 977–987 (2007)
95. Patel, G.K., Dosen, S., Castellini, C., Farina, D.: Multichannel electrotactile feedback for simultaneous and proportional myoelectric control. J. Neural Eng. **13**(5), 056,015 (2016)
96. Kim, K., Colgate, J.E.: Haptic feedback enhances grip force control of semg-controlled prosthetic hands in targeted reinnervation amputees. IEEE Trans. Neural Syst. Rehabil. Eng. **20**(6), 798–805 (2012)
97. Christ, O., Beckerle, P., Preller, J., Jokisch, M., Rinderknecht, S., Wojtusch, J., von Stryk, O., Vogt, J.: The rubber hand illusion: maintaining factors and a new perspective in rehabilitation and biomedical engineering? Biomed. Eng. **57**(S1), 1098–1101 (2012)
98. Bicchi, A., Gabiccini, M., Santello, M.: Modelling natural and artificial hands with synergies. Philos. Trans. R. Soc. B **366**(1581), 3153–3161 (2011)
99. Hayward, V.: Is there a 'plenhaptic' function? Philos. Trans. R. Soc. B **366**(1581), 3115–3122 (2011)
100. Bianchi, M., Serio, A.: Design and characterization of a fabric-based softness display. IEEE Trans. Haptics **8**(2), 152–163 (2015)
101. Bianchi, M., Valenza, G., Lanata, A., Greco, A., Nardelli, M., Bicchi, A., Scilingo, E.P.: On the role of affective properties in hedonic and discriminant haptic systems. Int. J. Soc. Robot. 1–9 (2016)
102. Bianchi, M., Valenza, G., Greco, A., Nardelli, M., Battaglia, E., Bicchi, A., Scilingo, E.: Towards a novel generation of haptic and robotic interfaces: integrating affective physiology in human-robot interaction. In: IEEE International Symposium on Robot and Human Interactive Communication, pp. 125–131. IEEE (2016)
103. Lo, A.C., Guarino, P.D., Richards, L.G., Haselkorn, J.K., Wittenberg, G.F., Federman, D.G., Ringer, R.J., Wagner, T.H., Krebs, H.I., Volpe, B.T., Bever, C.T., Bravata, D.M., Duncan, P.W.,

Corn, B.H., Maffucci, A.D., Nadeau, S.E., Conroy, S.S., Powell, J.M., Huang, G.D., Peduzzi, P.: Robot-assisted therapy for long-term upper-limb impairment after stroke. N. Engl. J. Med. **362**(19), 1772–1783 (2010)

104. Torricelli, D., Rodriguez-Guerrero, C., Veneman, J.F., Crea, S., Briem, K., Lenggenhager, B., Beckerle, P.: Benchmarking wearable robots: challenges and recommendations from functional, userexperience and methodological perspectives (in revision)
105. Zhang, J., Fiers, P., Witte, K.A., Jackson, R.W., Poggensee, K.L., Atkeson, C.G., Collins, S.H.: Human-in-the-loop optimization of exoskeleton assistance during walking. Science **356**(6344), 1280–1284 (2017)
106. Schültje, F., Beckerle, P., Grimmer, M., Wojtusch, J., Rinderknecht, S.: Comparison of trajectory generation methods for a human-robot interface based on motion tracking in the Int^{2}Bot. In: IEEE International Symposium on Robot and Human Interactive Communication (2014)
107. De Beir, A., Caspar, E.A., Yernaux, F., Magalhães Da Saldanha da Gama, P.A., Vanderborght, B., Cleermans, A.: Developing new frontiers in the rubber hand illusion: design of an open source robotic hand to better understand prosthetics. In: IEEE International Symposium on Robot and Human Interactive Communication (2014)
108. Dummer, T., Picot-Annand, A., Neal, T., Moore, C.: Movement and the rubber hand illusion. Perception **38**, 271–280 (2009)
109. Kalckert, A., Ehrsson, H.H.: Moving a rubber hand that feels like your own: a dissociation of ownership and agency. Front. Hum. Neurosci. **6**(40) (2012)
110. Arata, J., Hattori, M., Ichikawa, S., Sakaguchi, M.: Robotically enhanced rubber hand illusion. IEEE Trans. Haptics (2014)
111. Hara, M., Nabae, H., Yamamoto, A., Higuchi, T.: A novel rubber hand illusion paradigm allowing active self-touch with variable force feedback controlled by a haptic device. IEEE Trans. Hum. Mach. Syst. **46**(1), 78–87 (2016)
112. Beckerle, P., Christ, O., Wojtusch, J., Schuy, J., Wolff, K., Rinderknecht, S., Vogt, J., von Stryk, O.: Design and control of a robot for the assessment of psychological factors in prosthetic development. In: IEEE International Conference on Systems, Man and Cybernetics (2012)
113. Penner, D., Abrams, A.M.H., Overath, P., Vogt, J., Beckerle, P.: Robotic leg illusion: system design and human-in-the-loop evaluation. IEEE Trans. Hum. Mach. Syst. (2019)
114. Beckerle, P., Bianchi, M., Castellini, C., Salvietti, G.: Mechatronic designs for a robotic hand to explore human body experience and sensory-motor skills: a Delphi study. Adv. Robot. **32**(12), 670–680 (2018)
115. Abrams, A.M.H., Beckerle, P.: A pilot study: advances in robotic hand illusion and its subjective experience. In: HRI Pioneers 2018 (ACM/IEEE International Conference on Human-Robot Interaction), pp. 289–290 (2018)
116. Beckerle, P., Schültje, F., Wojtusch, J., Christ, O.: Implementation, control and user-feedback of the Int^{2}Bot for the investigation of lower limb body schema integration. In: IEEE International Symposium on Robot and Human Interactive Communication (2014)
117. Schürmann, T., Overath, P., Christ, O., Vogt, J., Beckerle, P.: Exploration of lower limb body schema integration with respect to body-proximal robotics. In: IEEE International Forum on Research and Technologies for Society and Industry Leveraging a Better Tomorrow (2015)
118. Lloyd, D.M.: Spatial limits on referred touch to an alien limb may reflect boundaries of visuo-tactile peripersonal space surrounding the hand. Brain Cogn. **64**, 104–109 (2007)
119. Preston, C.: The role of distance from the body and distance from the real hand in ownership and disownership during the rubber hand illusion. Acta Psychol. **142**(2), 177–183 (2013)
120. Constantini, M., Haggard, P.: The rubber hand illusion: sensitivity and reference frame for body ownership. Conscious. Cogn. **16**(2), 229–240 (2007)
121. Haans, A., IJsselsteijn, W.A., de Kort, Y.A.W.: The effect of similarities in skin texture and hand shape on perceived ownership of a fake limb. Body Image **5**(4), 389–394 (2008)
122. Tsakiris, M., Longo, M.R., Haggard, P.: Having a body versus moving your body: neural signatures of agency and body-ownership. Neuropsychologia **48**, 2740–2749 (2010)
123. Choi, W., Li, L., Satoh, S., Hachimura, K.: Multisensory integration in the virtual hand illusion with active movement. BioMed Res. Int. **2016** (2016)

124. Escamilla, R.F.: Knee biomechanics of the dynamic squat exercise. Med. Sci. Sports Exerc. **33**, 127–141 (2001)
125. Christ, O., Weber, C., Borchert, I., Sorgatz, H.: Dissonance between visual and proprioceptive information as a moderator in experimental pain. J. Rehabil. Med. **43**(9), 836 (2011)
126. Kalckert, A., Ehrsson, H.H.: The onset time of the ownership sensation in the moving rubber hand illusion. Front. Psychol. **8**, 344 (2017)
127. De Vignemont, F.: Embodiment, ownership and disownership. Conscious. Cogn. **20**(1), 82–93 (2011)

Part II
Upper Limbs

Chapter 3
Robotic Hand Experience

Abstract Augmenting rubber hand illusion paradigms extends experimental possibilities and can disentangle human action-perception loops. This chapter presents technical and experimental merits with a focus on tactile feedback and the influence of delays. A pilot study proofs the applicability of the RobHI setup with tactile feedback. Additionally, two extensive experiments analyze the interplay of visuotactile and visuomotor feedback to unravel their contributions to spatial body representation. Both vary motor and tactile feedback availability and delays while measuring the spatial calibration of body coordinates and subjective embodiment experience. The results indicate that both feedback types contribute similarly to embodiment and, interestingly, show that synchronous feedback in one factor can even compensate for asynchronous information from the other. These fundamental insights on multisensory enhancing effects confirm human-in-the-loop experiment potentials.

3.1 Upper Limb Body Experience

As discussed in Chap. 2, bodily self-experience relies on simultaneous processing and combination of various sensorimotor signals that contribute to a coherent representation of the body [1]. Body representations appear to be connected to representations of the reachable space surrounding the body, i.e., the peripersonal space [2], which facilitates humans to distinguish their bodies from the environment and enables them to execute movements and to interact [3]. Moreover, body and peripersonal space representations exhibit a intensively researched plasticity, which makes humans capable of using tools and integrating external artifacts into their body schemas [4].

With the rubber hand illusion [5] and its various implementations, research has developed and is still improving experimental paradigms to explore this plasticity. Synchronous sensory stimulation has been shown to be an important modulator of the observed embodiment effect since asynchronous signals dissolve the illusory sensation [6, 7]. The proprioceptive drift that is used as a behavioral proxy of embodiment is often seen as a consequence of a multimodal recalibration process of body representation within peripersonal space [6, 8]. With the advent of human-in-the-loop experiments supported by robotic, haptic, and virtual reality technologies,

P. Beckerle, *Human-Robot Body Experience*, Springer Series on Touch and Haptic Systems, https://doi.org/10.1007/978-3-030-38688-7_3

paradigms have been significantly extended over passive setups with stimulation through the experimenter [9–11]. As movement provides essential information for body and peripersonal space perception, moving artificial limbs were introduced: it is not yet fully answered whether active and passive movements induce ownership equally [12, 13], while only active movements calibrate the spatial body representation [14]. Remarkably, the moving rubber hand illusion is similarly weaker in case of asynchronous visuomotor stimulation [15, 16]. With robotic hand illusion (RobHI) experiments, experimental possibilities are further improved and extended [17–20].

To investigate the influence of haptic interfaces, Abrams et al. [21] and Huynh et al. [22] introduced vibrotactile feedback to the rubber hand illusion paradigm. Both works are subsequently presented to outline how human-in-the-loop techniques can extend experimental possibilities, e.g., by concurrently examining influences of visuotactile and visuomotor feedback. In contrast to previous RobHI experiments, the setups applied in both studies do not only precisely control visuomotor feedback (immediate or delayed) but facilitate independent manipulation of tactile feedback timing (immediate or delayed). Thereby, the setups allow to disentangle both effects like it would not be possible in classic experiments [22].

3.2 Robotic Hand Illusion (RobHI) with Tactile Feedback

Abrams et al. [21] performed a pilot study to demonstrate the advances that tactile feedback can add to the robotic hand illusion. To this end, they used a copy of the robotic hand introduced previously by Caspar, De Beir et al. [18, 23] equipped with tactile user feedback [21, 24]. Feedback is realized by measuring contact pressure on the robot's fingertips and mapping it to the human hand using vibration motors. Data acquisition, motion control, and feedback generation are implemented on the microcontroller.

Experimentally, Abrams et al. [21] examined different activity levels regarding their influence on embodiment: a static setup replacing the rubber hand by the robotic hand (Fig. 3.1a), the moving robotic hand without feedback (Fig. 3.1b), and a scenario including active motion and applying feedback (Fig. 3.1c). Conducting three experiments to evaluate the abilities of the robotic hand to reliably induce an illusion, proprioceptive drift is measured as a behavioral correlate of embodiment. The participants indicated the perceived localization deviation with eyes closed by pointing to where they experienced their right hand with the left index finger. In the study, 70 right-handed participants (78.6% female, 21.4% male; average age of 22.3 years, standard deviation of 3.9 years) were assigned to three groups conducting one experimental scenario each. External auditory influences were reduced by providing all participants with headphones dampening ambient sounds. The study was approved by the local ethics committee of the Department of Psychology at Technical University Darmstadt, Germany, and conducted in accordance with the Declaration of Helsinki.

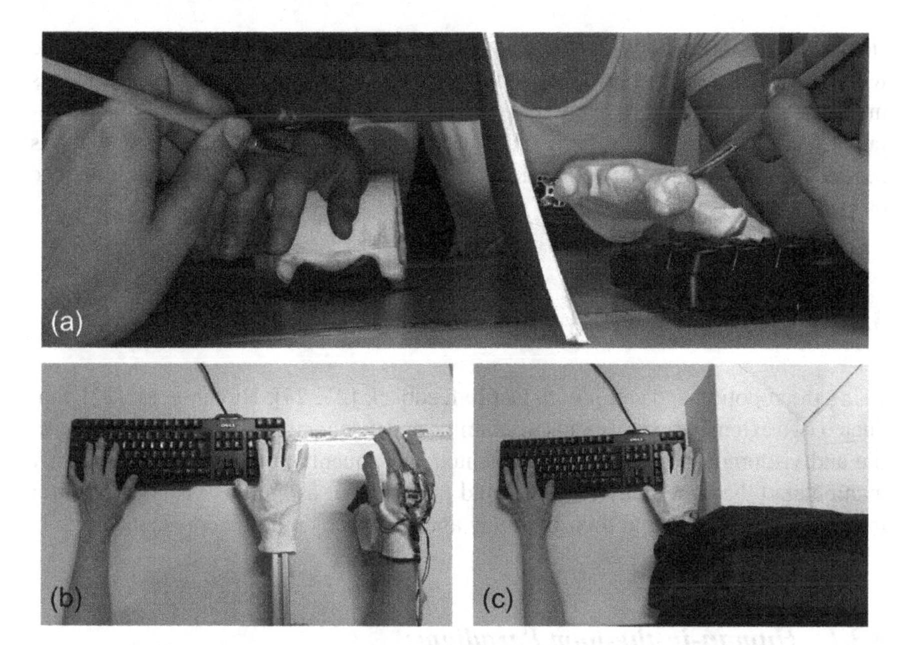

Fig. 3.1 Three experimental scenarios investigated by Abrams et al. [21]: Robot hand illusion without motion and with manual tactile stimulation (**a**), moving robot hand illusion (**b**), and moving robot hand illusion with tactile feedback. *Based on figures from* Abrams et al. [21]

Aiming at a replication of the classic rubber hand illusion [5], the proximal to the distal phalanxes of the right index fingers of the real and the robotic were tactilely stimulated with continuous manual brushstrokes in the first experiment. After 6 min of stimulation at a frequency of 0.5 Hz, a significant difference between pre-treatment and post-treatment proprioceptive drift was observed ($t_{21} = 2.52$, $p = 0.02 < 0.05$). In the second experiment with active movement, Abrams et al. [21] replicated the experiment of Caspar et al. [18] by including the active movement of the robotic hand controlled by the participants without perceptible delay. Participants moved their right index and middle fingers up and down in temporal coherence with a metronome rhythm on their headphones to control the robotic hand stroking a computer keyboard. Statistical analysis of proprioceptive drift data outlines significant results ($t(23) = 3.28$, $p = 0.003 < 0.05$) and confirms a replication of the previous research. Finally, the third experiment went beyond previous studies by adding vibrotactile feedback to provide the participants with contact information. As was expected, a significant difference between pre- and post-experimental proprioceptive drift measurements was found ($t(23) = 4.83$, $p < 0.001$) underlining the elicitation of the bodily illusion.

With their study, Abrams et al. [21] highlighted the feasibility of extending robotic hand illusion experiment with tactile feedback. Enabling new experimental possibilities, they envisioned future studies to extent embodiment research and the consideration of human body experience in wearable robot design focusing on haptic feedback. Moreover, the perceived autonomy of humans using assistive devices is mentioned to be highly relevant and related to the agency as an embodiment subfactor (see Chap. 4). Accordingly, they suggest future research to analyze relations between control by the user and the reception of sensory feedback and their influence on bodily experience.

3.3 Feedback Interplay in the RobHI

Using the robotic hand setup with tactile feedback [21, 24], Huynh et al. [22] conducted two extensive experiments to examine and disentangle influences of visuotactile and visuomotor feedback. As presented subsequently, the study independently manipulated the synchrony of visual and tactile input as well as visual and motor information to understand their interplay and influence on body representation.

3.3.1 Human-in-the-loop Paradigms

The human-in-the-loop setup of Huynh et al. [22] is presented in Fig. 3.2. In the first experiment, data was obtained from 44 persons (52.3% female, 47.7% male; average age of 21.0 years, standard deviation of 2.3 years). The second experiment is based on trials with 18 newly recruited participants (38.9% female, 61.1% male; average age of 22.7 years, standard deviation of 2.0 years). The investigated sample size was selected taking into account expected effect sizes for the first experiment while the second sample was adjusted to the larger-than-expected effect sizes observed in this first experiment. The participants were naive to the experimental design, took part voluntarily, and gave informed consent during instruction. The study was conducted considering recommendations of the ethics committee of Technical University Darmstadt, Germany, and in accordance with the Declaration of Helsinki.

Participants were seated in front of a table and wearing passive noise-canceling headphones. Their investigated hands were hidden and parallely aligned with the visible robotic hand on the tabletop (see Fig. 3.2). The distance between the real and the robot hand was set to 21.0 cm to be in accordance with the spatial constraints of the rubber hand illusion [25, 26]. The custom-built robotic hand consists of 3D-printed link elements, which are flexibly connected by steel springs [18, 23]. Implementing tactile feedback, contact information is measured through pressure sensors installed on the robotic fingertips and mapped to vibration motors placed on the fingertips of the participants [21, 22, 24]. Human hand motions are acquired via flex sensors attached to each finger and mapped to the underactuated robot fingers by five servo motors and nylon cords. The system-intrinsic delay of motion mapping was quantified to be roughly 120 ms, while the vibrotactile mapping was found to exhibit an averaged delay of about 80 ms. During the interaction, participants were instructed

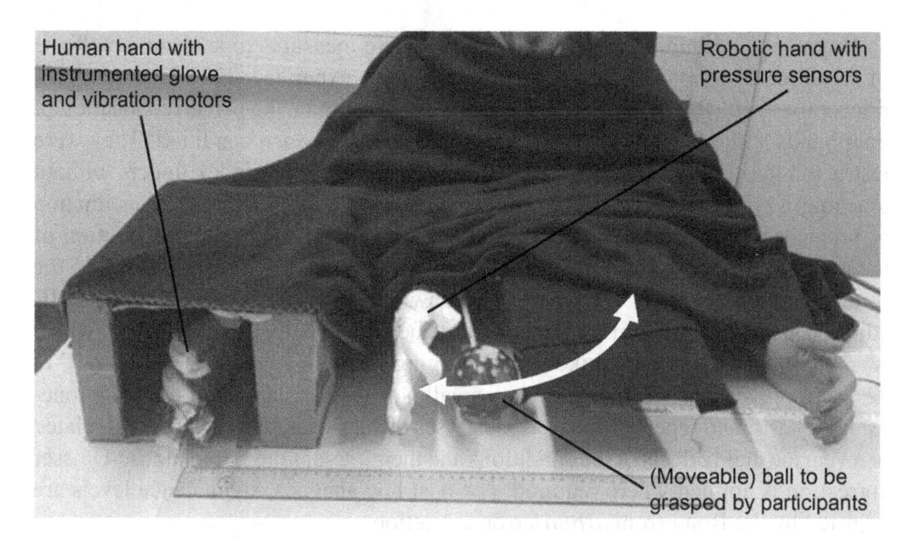

Fig. 3.2 Experimental setup used by Huynh et al. [22]: the hand of the participant is hidden and equipped with an instrumented glove that acquires motion data and provides vibrotactile feedback. The visible robotic hand mimics the grasping motions of the human and measures contact pressures. For the second experiment, the ball to be grasped is attached to a stick on a motor able to move it on a circular trajectory. *Based on a figure from* Huynh et al. [22]

to repeatedly grasp a ball with the robotic hand, which they controlled through their hand and the sensorized glove (see Fig. 3.2). For their second experiment, Huynh et al. attached this ball to a motor via a stick enabling it to move on a circular trajectory [22].

At first, Huynh et al. conducted a $2x2$ repeated-measures experiment manipulating tactile feedback synchrony (synchronous = TFs vs. asynchronous = TFa) and motor feedback synchrony (synchronous = MFs vs. asynchronous = MFa). The four conditions were presented during interaction phases of 180 s in which the participants grasped the statically located ball. A Latin square was used to avoid sequence effects. In the asynchronous conditions, an additional delay of 500 ms was added to the system-intrinsic values through reprogramming timers on the microcontroller commanding the motors.

After systematically varying synchrony of tactile and motor feedback, Hunyh et al. added unimodal conditions to probe the interaction of synchronous and asynchronous modalities [22]. To this end, they implemented a $3x3$ repeated-measures experiment manipulating tactile (synchronous = TFs vs. asynchronous = TFa vs. no feedback = TFn) and motor feedback (synchronous = MFs vs. asynchronous = MFa vs. no feedback = MFn) using a Latin square. To realize the unimodal TF conditions (MFn/TFs and MFn/TFa), the ball was moved by a servo motor as indicated in Fig. 3.2 to stimulate the pressure sensors without requiring self-executed motions from the participants.

Huynh et al. combined subjective and objective measures to analyze the effects of the different conditions on the participants' body experience [22]. To assess proprioceptive drift (PPD), participants were asked to estimate the perceived location of their hands at the beginning of the experiment and after each condition. They were instructed to point the back of their right hand with their left index fingers without touching it while keeping their eyes closed. Aiming at a more accurate estimation of proprioceptive drift, participants were asked to blindly indicate the position of their real hand three times after each condition of the second experiment and the results were averaged [5, 22]. Each condition was concluded by surveying a RobHI-adjusted German version of the embodiment questionnaire by Longo et al. [7], which was answered on a 5-point Likert scale ranging from "−2 = strongly disagree" to "2 = strongly agree" (Table. 3.1). Statistically, differences are analyzed using one-sample t-tests, and repeated-measures analyses of variances (ANOVA) are calculated to reveal the main effects of the conditions and their interactions. Greenhouse-Geisser correction is applied where required. For post hoc analysis, significance levels are adjusted by the Bonferroni $alpha$-error correction.

Table 3.1 RobHI-adjusted version of the embodiment questionnaire by Longo et al. [7] and the version translated to German as used by Huynh et al. [22]

Item #	Original statement during the block...	Translated item Während der Durchführung...
1.	... it seemed like I was looking directly at my own hand, rather than a robotic hand	... hatte ich den Eindruck, direkt meine eigene Hand anzuschauen, nicht eine Roboterhand
2.	... it seemed like the robotic hand began to resemble my real hand	... schien es, als ob die Roboterhand meiner realen Hand ähnlicher wurde
3.	... it seemed like the robotic hand belonged to me	... schien es, als ob die Roboterhand zu mir gehören würde
4.	... it seemed like the robotic hand was my hand	... schien es, als ob die Roboterhand meine eigene Hand war
5.	... it seemed like the robotic hand was part of my body	... schien es, als ob die Roboterhand ein Teil meines Körpers war
6.	... it seemed like my hand was in the position where the robotic hand was	... schien es, als ob meine eigene Hand in der Position der Roboterhand war
7.	... it seemed like the robotic hand was in the position where my hand was	... schien es, als ob die Roboterhand in der Position meiner eigenen Hand war
8.	... it seemed like I could have moved the robotic hand if I had wanted	... hatte ich den Eindruck, dass ich die Roboterhand bewegen könnte, wenn ich gewollt hätte
9.	... it seemed like I was in control of the robotic hand	... hatte ich den Eindruck, die Kontrolle über die Roboterhand zu haben

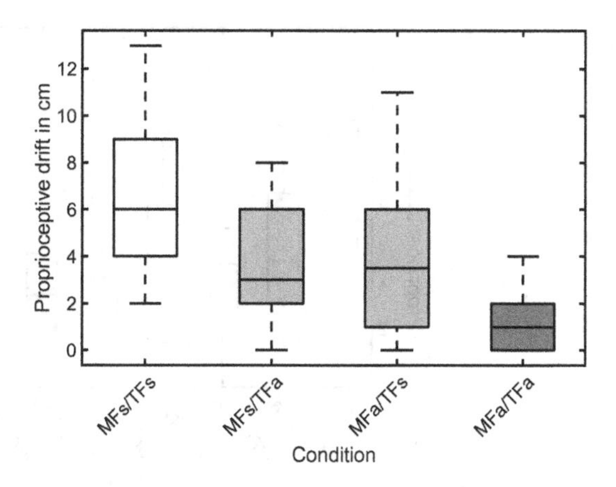

Fig. 3.3 Means and standard deviations of proprioceptive drift towards the robotic hand in the first experiment. Color codes synchrony: white—fully synchronous condition, light grey—mixed synchrony conditions, dark grey—fully asynchronous condition

3.3.2 Results and Observations

In their first experiment, Hunyh et al. [22] observed significant proprioceptive drifts in each condition (all $p < 0.001$, see [22] for details) with the highest result in the MFs/TFs condition ($M_{MFs/TFs} = 7.00\,\text{cm}$ $SD_{MFs/TFs} = 3.26\,\text{cm}$). Figure 3.3 presents the proprioceptive drift results underlining the effect of the fully synchronous condition and, furthermore, shows that the fully asynchronous condition yielded the lowest drifts. The ANOVA reveals a significant main effect for both motor feedback ($F_{1,43} = 168.18$, $p < 0.001$) and tactile feedback ($F_{1,43} = 104.99$, $p < .001$), but no significant interaction ($F_{1,43} = 1.04$, $p = 0.32$). Post hoc analysis confirmed a significant difference of proprioceptive drift ($p < 0.001$) between the synchronous and asynchronous conditions in both modalities. While this outlines that motor and tactile feedback individually influence perceived hand position without interacting with each other, contrast testing MFa/TFs against MFs/TFa suggests that synchrony of one factor might compensate for asynchrony of the other [22]. An ANOVA of the averaged questionnaire data shows a significant main effect of motor feedback ($F_{1,43} = 20.22$, $p < 0.001$), with post hoc tests revealing higher ratings in the synchronous compared to the asynchronous levels ($p < 0.001$). In contrast to proprioceptive drift results, no significant main effect of tactile feedback on subjective embodiment is found ($F_{1,43} = 2.03$, $p = 0.16$). The absence of interaction between the factors is underlined by the subjective results ($F_{1,43} = 0.89$, $p = 0.35$).

As their first experiment, the second experiment of Hunyh et al. [22] yielded significant proprioceptive drifts in each condition (all $p < 0.001$, see [22] for details). Among the results, which are shown in Fig. 3.4, the highest proprioceptive drift is observed in the MFs/TFs condition again ($M_{MFs/TFs} = 9.01\,\text{cm}$ $SD_{MFs/TFs} = 3.61\,\text{cm}$). The repeated-measures ANOVA of all conditions reveals significant main effects of motor feedback ($F_{1.51,25.59} = 106.80$, $p < 0.001$) and tactile feedback ($F_{2,34} = 69.87$, $p < 0.001$) as well as a significant interaction between both factors ($F_{2.84,48.19} = 20.34$, $p < 0.001$). Post hoc factor level testing reveals that

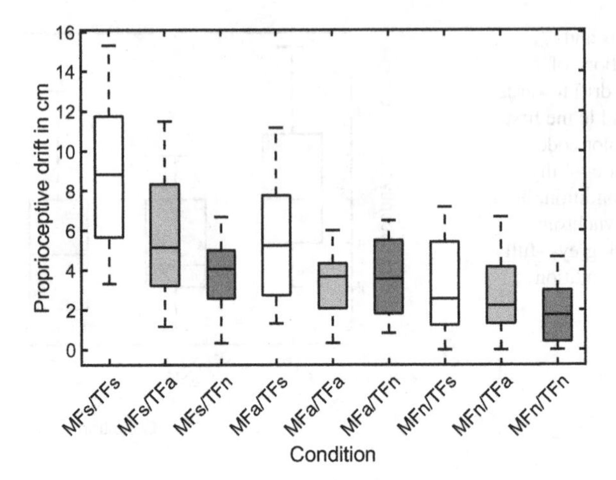

Fig. 3.4 Means and standard deviations of proprioceptive drift towards the robotic hand in the second experiment. Colors code conditions of tactile feedback: white—synchronous, light grey—asynchronous, dark grey—no feedback

effects decreased from the synchronous to the asynchronous and, further, to the no feedback condition for both motor feedback (all $p < 0.001$) and tactile feedback (all $p \leq 0.002$). This leads to the conclusion that asynchronous feedback information seems to be exploited at least partly for the calibration of spatial body coordinates (see MFs/TFa and MFa/TFs conditions in Fig. 3.4). Further post hoc comparisons focusing on synchronous and asynchronous conditions confirm the replication of the findings of the first experiment. Analyzing subjective data from the embodiment questionnaire reveals a significant main effect of tactile feedback ($F_{1.39,23.54} = 14.24$, $p < 0.001$), but no significant effect of motor feedback ($F_{1.35,22.98} = 2.80$, $p = 0.1$) and no interaction between the factors ($F_{3.15,53.60} = 0.51$, $p = 0.69$).

3.4 Discussion and Perspectives

The outcome of Abrams et al. [21] outlines the feasibility of robotic hand illusion experiments and the potential of extending these human-in-the-loop paradigms with tactile feedback. The results on the moving RobHI by Caspar et al. [18] are replicated and, moreover, Abrams et al. [21] demonstrate successful embodiment in a tactilely enhanced moving RobHI for the first time, which is concluded to open up a variety of future experiments.

Picking up on this, the extensive study of Hunyh et al. [22] dug deeper into the influences and interactions of motor and tactile feedback in two experiments using the human-in-the-loop approach to independently manipulate visuomotor and visuotactile feedback. From their first experiment, Hunyh et al. [22] conclude that synchronous motor feedback as well as synchronous tactile feedback induced significantly higher proprioceptive drifts towards the robotic hand. Remarkably, their results imply that synchrony of one kind of feedback can compensate for the asynchrony of the other regarding the calibration of spatial body coordinates. The second experiment of Huynh et al. [22] replicated most of their previous findings. Through the consideration of unimodal conditions, the results further suggest that fusing information from both kinds of feedback results in a super-additive effect on the calibration of body-space coordinates. Interpreting the results of both experiments, Hunyh et al. furthermore suggest an enhancing effect on body-space representation through sensorimotor combination and integration of multimodal feedback.

Generally, the findings of Hunyh et al. [22] regarding synchronous and asynchronous stimulation are in accordance with previous rubber hand illusion studies [7, 27, 28], which underlines the feasibility of the robot hand illusion. As Rohde et al. [27], proprioceptive drift occurred even in asynchronous stimulation conditions as long as a visuoproprioceptive conflict is induced. Moreover, their results replicate the early RobHI demonstration of Romano et al. [29], which showed the importance of visuomotor correlations, and additionally show the relevance of its interaction with visuotactile feedback. While Romano et al. did not observe any robotic hand embodiment experiences, the participants of Huynh et al. reported embodiment experiences, but data is rather inconsistent [22, 29]. Since the observed effects were rather small compared to observed proprioceptive drifts, engineers might be advised to reconsider design requirements such as adapting the appearance of the robotic hand [24, 30] or matching the participants' anthropometry [24, 31].

Overall, the three experiments discussed in this chapter demonstrate the potential of tactilely enhanced robotic hand illusions studies. Huynh et al. demonstrate that such human-in-the-loop experiments can provide implications for the understanding of multimodal mechanisms of body and peripersonal space representations going beyond classical experimental possibilities, e.g., by decomposing effects that are inseparably connected in humans [22]. Observations like the apparently equal contribution of visuotactile and visuomotor feedback to implicit measures of spatial body representation can provide suggestions for multimodal human-machine interface design [22, 32]. Therefore, future studies might vary the implementation of feedback, e.g., with respect to modality (see Sect. 4.1) or location. Moreover, all three experiments are paramount examples of how human-in-the-loop experiments might directly guide the design of wearable and assistive robotic systems by examining how people respond to real hardware, which might be limited in virtual reality experiments.

References

1. Tsakiris, M.: My body in the brain: a neurocognitive model of body-ownership. Neuropsychologia **48**(3), 703–712 (2010)
2. Serino, A., Noel, J.P., Galli, G., Canzoneri, E., Marmaroli, P., Lissek, H., Blanke, O.: Body part-centered and full body-centered peripersonal space representations. Sci. Rep. **5**, 18,603 (2015)
3. Kalveram, K.T.: Wie das Individuum mit seiner Umwelt interagiert: psychologische, biologische und kybernetische Betrachtungen über die Funktion von Verhalten. Pabst Science (1998)
4. Holmes, N.P., Spence, C.: The body schema and the multisensory representation(s) of peripersonal space. Cogn. Process. **5**(2), 94–105 (2004)
5. Botvinick, M., Cohen, J.: Rubber hands 'feel' touch that eyes see. Nature **391**, 756 (1998)
6. Tsakiris, M., Haggard, P.: The rubber hand illusion revisited: visuotactile integration and self-attribution. J. Exp. Psychol.: Hum. Percept. Perform. **31**(1), 80–91 (2005)
7. Longo, M.R., Schüür, F., Kammers, M.P.M., Tsakiris, M., Haggard, P.: What is embodiment? A psychometric approach. Cognition **107**, 978–998 (2008)
8. Christ, O., Reiner, M.: Perspectives and possible applications of the rubber hand and virtual hand illusion in non-invasive rehabilitation: technological improvements and their consequences. Neurosci. Biobehav. Rev. **44**, 33–44 (2014)
9. Slater, M., Pérez Marcos, D., Ehrsson, H., Sanchez-Vives, M.V.: Towards a digital body: the virtual arm illusion. Front. Hum. Neurosci. **2**(6) (2008)
10. Rosen, B., Ehrsson, H.H., Antfolk, C., Cipriani, C., Sebelius, F., Lundborg, G.: Referral of sensation to an advanced humanoid robotic hand prosthesis. Scand. J. Plast. Reconstr. Surg. Hand Surg. **43**, 260–266 (2009)
11. Bekrater-Bodmann, R., Foell, J., Diers, M., Kamping, S., Rance, M., Kirsch, P., Trojan, J., Fuchs, X., Bach, F., Cakmak, H.K., Maaß, H., Flor, H.: The importance of synchrony and temporal order of visual and tactile input for illusory limb ownership experiences—an FMRI study applying virtual reality. PLoS One **9**(1), e87,013 (2014)
12. Raz, L., Weiss, P.L., Reiner, M.: The virtual hand illusion and body ownership. In: International Conference on Human Haptic Sensing and Touch Enabled Computer Applications, pp. 367–372. Springer (2008)
13. Dummer, T., Picot-Annand, A., Neal, T., Moore, C.: Movement and the rubber hand illusion. Perception **38**, 271–280 (2009)
14. Tsakiris, M., Prabhu, G., Haggard, P.: Having a body versus moving your body: how agency structures body-ownership. Conscious. Cogn. **15**(2), 423–432 (2006)
15. Kalckert, A., Ehrsson, H.H.: Moving a rubber hand that feels like your own: a dissociation of ownership and agency. Front. Hum. Neurosci. **6**(40) (2012)
16. Sanchez-Vives, M.V., Spanlang, B., Frisoli, A., Bergamasco, M., Slater, M.: Virtual hand illusion induced by visuomotor correlations. PloS One **5**(4), e10,381 (2010)
17. Arata, J., Hattori, M., Ichikawa, S., Sakaguchi, M.: Robotically enhanced rubber hand illusion. IEEE Trans. Haptics (2014)
18. Caspar, E.A., de Beir, A., Magalhães Da Saldanha da Gama, P.A., Yernaux, F., Cleeremans, A., Vanderborght, B.: New frontiers in the rubber hand experiment: when a robotic hand becomes one's own. Behav. Res. Methods **47**(3), 744–755 (2015)
19. Hara, M., Nabae, H., Yamamoto, A., Higuchi, T.: A novel rubber hand illusion paradigm allowing active self-touch with variable force feedback controlled by a haptic device. IEEE Trans. Hum. Mach. Syst. **46**(1), 78–87 (2016)
20. Ismail, M.A.F., Shimada, S.: Robot hand illusion under delayed visual feedback: relationship between the senses of ownership and agency. PLOS ONE **11**(7), e0159,619 (2016)
21. Abrams, A.M.H., Beckerle, P.: A pilot study: advances in robotic hand illusion and its subjective experience. In: HRI Pioneers 2018 (ACM/IEEE International Conference on Human-Robot Interaction), pp. 289–290 (2018)

22. Huynh, T.V., Bekrater-Bodmann, R., Fröhner, J., Vogt, J., Beckerle, P.: Robotic hand illusion with tactile feedback: unravelling the relative contribution of visuotactile and visuomotor input to the representation of body parts in space. PloS One **14**(1), e0210,058 (2019)
23. De Beir, A., Caspar, E.A., Yernaux, F., Magalhães Da Saldanha da Gama, P.A., Vanderborght, B., Cleermans, A.: Developing new frontiers in the rubber hand illusion: design of an open source robotic hand to better understand prosthetics. In: IEEE International Symposium on Robot and Human Interactive Communication (2014)
24. Beckerle, P., De Beir, A., Schürmann, T., Caspar, E.A.: Human body schema exploration: analyzing design requirements of robotic hand and leg illusions. In: IEEE International Symposium on Robot and Human Interactive Communication (2016)
25. Lloyd, D.M.: Spatial limits on referred touch to an alien limb may reflect boundaries of visuo-tactile peripersonal space surrounding the hand. Brain Cogn. **64**, 104–109 (2007)
26. Kalckert, A., Ehrsson, H.H.: The moving rubber hand illusion revisited: comparing movements and visuotactile stimulation to induce illusory ownership. Conscious. Cogn. **26**, 117–132 (2014)
27. Rohde, M., Di Luca, M., Ernst, M.O.: The rubber hand illusion: feeling of ownership and proprioceptive drift do not go hand in hand. PLoS ONE **6**(6) (2011)
28. Flögel, M., Kalveram, K.T., Christ, O., Vogt, J.: Application of the rubber hand illusion paradigm: comparison between upper and lower limbs. Psychol. Res. **80**(2), 298–306 (2015)
29. Romano, R., Caffa, E., Hernandez-Arieta, A., Brugger, P., Maravita, A.: The robot hand illusion: inducing proprioceptive drift through visuo-motor congruency. Neuropsychologia **70**, 414–420 (2015)
30. Tsakiris, M., Carpenter, L., James, D., Fotopoulou, A.: Hands only illusion: multisensory integration elicits sense of ownership for body parts but not for non-corporeal objects. Exp. Brain Res. **204**(3), 343–352 (2010)
31. Pavani, F., Zampini, M.: The role of hand size in the fake-hand illusion paradigm. Perception **36**(10), 1547–1554 (2007)
32. Nostadt, N., Abbink, D.A., Christ, O., Beckerle, P.: Embodiment, presence, and their intersections: teleoperation and beyond (submitted). ACM Trans. Hum. Robot. Interact. (2020)

Chapter 4
Virtual Hand Experience

Abstract Similar to robotic approaches, virtual hand illusion (VHI) experiments enable deeper insights into human-robot body experience. This chapter discusses the influence of different haptic feedback modalities as well as interrelations to autonomy and controllability. The first study compares wearable force feedback, vibrotactile feedback, and no haptic feedback during pick-and-place tasks in a virtual environment. Both kinds of haptic feedback significantly improve the embodiment of the virtual hand and the human-in-the-loop paradigms can guide wearable haptic designs. A second line of research analyzes the physical interaction of humans with intelligent robotic tools in an immersive virtual reality. Results suggest embodiment as a valid metric for the user experience of shared control quality and, furthermore, agency as an objective measure of task-appropriate and intuitive assistance. This underlines the potential of technically augmented psychological paradigms, to investigate human-in-the-loop control of physical human-robot interaction.

4.1 Feedback Modalities in the Virtual Hand Illusion (VHI)

Haptic interaction devices have strong potential to augment virtual reality (VR) developments to improve the users' experiences with respect to presence and potentially embodiment [1–3], which yields increased research activity [4, 5] and force feedback mechanisms. Especially, the field of wearable haptic feedback commonly implemented by moving platforms, pin-arrays, shearing belts and tactors, pneumatic jets, and balloon-based systems receives increasing popularity [6].

Despite striking advances in haptic technology [6], most human body experience research relies on haptic feedback either constrained to the end effector manipulated by the user [7, 8] or using mechanically controlled exoskeletons [9]. Only recently, psychological research attempts to exploit the increasing possibilities emerging through wearable haptics [10–12]. Considering a study by Fröhner et al. [13], this section shows how human-in-the-loop approaches augmented with wearable haptic devices help to understand and assess the effect of different feedback modalities on human-robot body experience. The study compared haptic feedback modalities provided to users during completing a task in a virtual environment with respect to the

P. Beckerle, *Human-Robot Body Experience*, Springer Series on Touch and Haptic Systems, https://doi.org/10.1007/978-3-030-38688-7_4

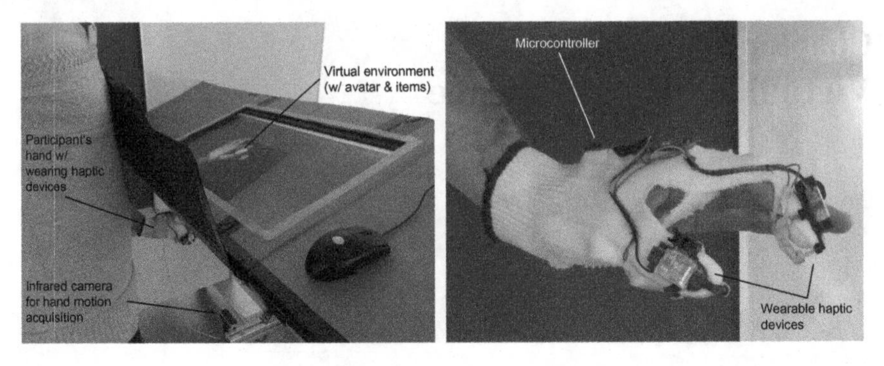

Fig. 4.1 The VHI experimental setup used by Fröhner et al. [13]: participants conduct a pick-and-place task moving items in a dynamic virtual environment (left) while being provided with vibratory and normal force feedback by the wearable haptic devices (right)

users' body experience. Hypothesizing VHI embodiment experience increases with improving haptic feedback fidelity, wearable force feedback [14] and vibrotactile stimulation are juxtaposed to each other and to providing no haptic feedback [13].

4.1.1 VHI Augmented by Wearable Haptics

Wearable haptic devices facilitate free movement during VHI experiments and enable focusing on visuomotor and visuotactile effects. Aiming to probe whether haptic feedback implementations influence VHI intensity, Fröhner et al. [13] used the setup presented left in Fig. 4.1 and investigated 32 persons participated (34.4% female, 65.6% male; 3 left-/ 29 right-handed) with an average age of 27.2 years and an age standard deviation of 7.2 years. The participants took part voluntarily, were naive to the particular experimental design, reported to have no perceptual constraints, and gave informed consent during instruction. The study was conducted considering the ethical guidelines of the European project WEARHAP coordinated by the University of Siena, Italy.

Participants were instructed to stand in front of a display and their left hand was equipped with a glove carrying the haptic system and allowing free motions as well as quick donning and doffing. Their task was to pick-and-place items to moving targets in the virtual environment. During the interaction, index finger and thumb could be stimulated each through vibration exerted by an eccentric mass motor and/or normal forces generated via platform driven by a servo motor as shown right in Fig. 4.1. The left hand was placed underneath the screen during interaction and hidden from the users' views by the display and cardboard covers, which were aligned with the body dimensions of the participants. To mirror the users' motions in the virtual environment, a sensor system with two monochromatic infrared cameras and three infrared LEDs under the screen acquired hand and finger motion data [13]. For fast visual feedback of the virtual avatar and environment, a gaming graphics card and a 144 Hz gaming display were used. A measurement of the system-intrinsic delay by Fröhner et al. [13] yielded a duration of 78.46 ms with a standard deviation of 6.67 ms between finger motion and visual feedback. Through a computer mouse on their right, participants could complete the questionnaire about their subjectively perceived body experience and input information for the determination of objective measures. To avoid biases, the experimenter hid himself during conduction, participants wore passive noise-canceling headphones, and subjective as well as objective measures were automatically evaluated within the virtual environment.

In a $3x2$ repeated-measures experiment, contact between the virtual hand and the virtual items triggered discrete feedback at the fingertips with one of the independently examined modalities, i.e., vibrotactile, force, or no feedback. Besides feedback modality, synchrony was examined as an independent variable: while the synchronous conditions were only subject to the very low system-intrinsic delay, an additional delay of 350 ms was added in the asynchronous conditions according to previous studies by Ismail and Shimada [15]. Asynchronous haptic feedback was expected to decrease embodiment experience significantly [16, 17], whereas the effect was expected to increase with improving haptic fidelity.

To assess VHI intensity in the different conditions, Fröhner et al. [13] considered proprioceptive drift (PPD), subjective embodiment, and task performance. Proprioceptive drift, i.e., the difference of the actual and experienced hand positions, was acquired before the experiment (pre-measurement) and after each condition (post-measurement). To this end, the screen was set to black right after the interaction phase and the participants were asked to point to the perceived position of their left index finger with the mouse cursor. To assess the subjective experience of embodiment, participants subsequently completed a VHI-adjusted version of the questionnaire by Longo et al. [16] that is shown in Table 4.1. Using visual analog scales mapping "-3 = Strongly disagree"–"3 = Strongly agree" and starting in the middle position, the participants reported their body experience regarding ownership, location, and agency regarding the virtual hand. Moreover, the number of successfully placed cubes on a moving target was determined as a measure of task performance. To determine condition-related differences in the objective and subjective measures, two-way repeated-measures analyses of variances (ANOVA) are calculated, Greenhouse-Geisser correction is applied if necessary, and equivalent tests were conducted where required. Significance levels were adjusted by the Bonferroni $alpha$-error correction for each dependent variable separately. Additionally, the results of hypothesis testing were double-checked by calculating Bayes factors interpreted with the scale of Kass and Raftery [18].

4.1.2 Results and Observations

The results of Fröhner et al. [13] indicate that the proprioceptive drift was neither influenced by feedback modality ($F = 11.41$, $p = 0.24$) nor by delay ($F = 11.56$,

Table 4.1 VHI-adjusted version of the embodiment questionnaire by Longo et al. [16] in English and the version translated to German as used by Fröhner et al. [13]

Item #	Original statement during the block...	Translated item Während des letzten Blocks...
1.	... it seemed like I was looking directly at my own hand, rather than at a virtual hand	... hatte ich den Eindruck, direkt meine eigene Hand anzuschauen, nicht eine virtuelle Hand
2.	... it seemed like the virtual hand began to resemble my real hand	... schien es, als ob die virtuelle Hand meiner realen Hand ähnlicher wurde
3.	... it seemed like the virtual hand belonged to me	... schien es, als ob die virtuelle Hand zu mir gehören würde
4.	... it seemed like the virtual hand was my hand	... schien es, als ob die virtuelle Hand meine eigene Hand war
5.	... it seemed like the virtual hand was part of my body	... schien es, als ob die virtuelle Hand ein Teil meines Körpers war
6.	... it seemed like my hand was in the location where the virtual hand was	... schien es, als ob meine eigene Hand in der Position der virtuellen Hand war
7.	... it seemed like the virtual hand was in the location where my hand was	... schien es, als ob die künstliche Hand in der Position meiner eigenen Hand war
8.	... it seemed like the touch I felt was caused by touching the virtual cube	... hatte ich den Eindruck, dass die Berührung, die ich fühlte, durch das Berühren des virtuellen Würfels verursacht wurde
9.	... it seemed like I could have moved the virtual hand if I had wanted	... hatte ich den Eindruck, dass ich die virtuelle Hand bewegen könnte, wenn ich gewollt hätte
10.	... it seemed like I was in control of the virtual hand	... hatte ich den Eindruck, die Kontrolle über die virtuelle Hand zu haben

$p = 0.14$) or their interaction ($F = 3.14$, $p = 0.62$). This is underlined by the distinct standard deviations in all synchronous conditions shown in Fig. 4.2, post hoc comparison with one-tailed paired t-test, and Bayes factors ($B_{10,ppd,modality} = 0.24$, $B_{10,ppd,delay} = 0.36$, $B_{10,ppd,interaction} = 0.08$).

Subjective embodiment is represented by the mean rating of the ten questionnaire items presented for all synchronous conditions in Fig. 4.3. In contrast to proprioceptive drift, Fröhner et al. [13] found a significant main effect of modality ($F = 9.77$, $p = 0.001^{**}$) and delay ($F = 10.44$, $p = 0.003^{*}$) on subjective embodiment. No interaction effect is observed ($F = 0.31$, $p = 0.73$). Bayes factor analysis substantiates the significant results of the ANOVA regarding both main effects with strong to very strong results ($B_{10,emb,modality} = 390.8$, $B_{10,emb,delay} = 56.5$, $B_{10,emb,interaction} = 42288.3$). A post hoc one-tailed paired t-test with Bonferroni α-error correction, indicates significant results between "no feedback" and "force feedback" ($p = 0.002^{*}$), "no feedback" and "vibration feedback" ($p = 0.006^{*}$), but not between "force feedback" and "vibration feedback" ($p = 0.1$).

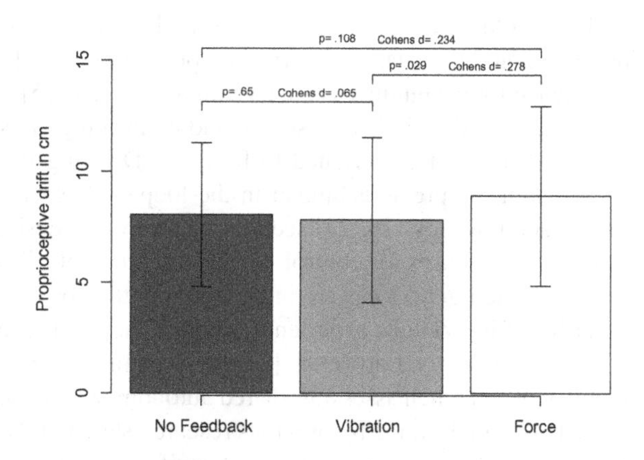

Fig. 4.2 Proprioceptive drift towards the virtual hand: means and standard deviations considering Bonferroni correction $\alpha = \frac{0.05}{3} = 0.017$. *Based on a figure from* Fröhner et al. [13]

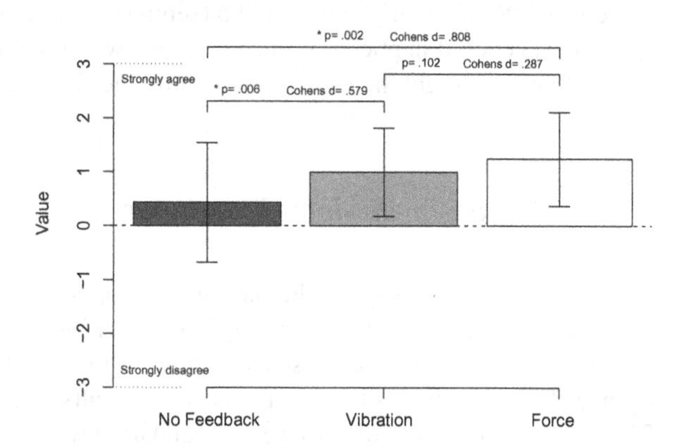

Fig. 4.3 Subjective embodiment surveyed by questionnaire: means and standard deviations considering Bonferroni correction $\alpha = \frac{0.05}{3} = 0.017$. *Based on a figure from* Fröhner et al. [13]

Moreover, Fröhner et al. [13] analyzed task performance observing no significant main effect of feedback modality ($p = 0.963$, $B_{10,perf,modality} = 0.06$), but higher performance without delay ($p = .001^{**}$, $B_{10,perf,delay} = 1326.7$).

4.2 Embodiment and Agency as Control Quality Metrics

The studies of Huynh et al. [19] and Fröhner et al. [13] outlined that human-in-the-loop experiments assessing embodiment could be of use in human-robot interface design. Subsequently, two studies by Fröhner et al. [20] and Endo et al. [21] started investigating whether the concepts of embodiment and agency could aid designing assistive controllers for human-robot interaction. This section summarizes the methods and results of these two studies to provide a holistic view of the possibilities and future developments in haptic interaction and (semi-)autonomous control.

To meet the users' demands in control design, explicit objectives are helpful for optimization, but usually performance-oriented [21–23]. Yet, the relevance of user experience for designing assistive technologies [24, 25] calls for methodologies for objective consideration, assessment, and monitoring of user experience in a holistic fashion that is not constrained to function. Drawing inspiration from psychological research and previous human-in-the-loop studies, embodiment [13, 16, 19] and its subfactor agency [26, 27] seem to be promising design metrics and, potentially, adaptation measures for control. While embodiment tries to capture how (artificial) limbs or (intelligent) tools are integrated into the body schema, the sense of agency describes if individuals experience whether they cause and control an action themselves. Accordingly, Endo et al. [21] focus on the sense of agency assuming that it is highly relevant when assessing shared autonomy tasks and directly links to intuitiveness. This is underlined by previous research showing that individuals might have a sense of agency over the actions of others if those comply sufficiently with their own objectives [28]. Similarly, a user and a (semi-)autonomous system might find some collective agency if haptic assistance is in line with the goals of the user. The user's sense of agency might then serve as a quality metric of a control system [21].

4.2.1 Human-in-the-loop Paradigms

Using different versions of a human-in-the-loop system with kinesthetic human-machine interaction in a virtual reality setting, Fröhner et al. [20] examine the suitability of embodiment and presence, while Endo et al. [21] focus on the sense of agency. The study by Fröhner et al. analyzed results of 8 participants (25% female, 75% male; average age of 26.1 years, standard deviation of 1.0 years). Endo et al. investigate data from 22 participants (13.6% female, 86.4% male; average age of 25.0 years, standard deviation of 3.0 years). Both studies rely on written, informed consent by all participants and were approved by the research ethics committee of Technical University of Munich, Germany.

In both studies, the participants physically interacted with a manipulandum with their right hand to control an avatar/cursor as shown in Fig. 4.4. Two perpendicularly stacked, linearly actuated carts provide kinesthetic rendering in a planar workspace. The handle hold by the participants is mounted on the upper cart and instrumented with a 6-axis force/torque sensor. Fröhner et al. [20] use head-mounted display to present a hand avatar in a virtual environment. This avatar is aligned with the right arm posture of the participants online through inverse kinematics calculations. Endo et al. [21] provide visual information with a computer monitor placed face-down over the manipulandum and mirrored to the users' field of view. The location of the cursor to be controlled was calibrated to match the center of the manipulandum handle.

Fröhner et al. [20] use a $2x2$ repeated-measures design controlling sequence effects through applying Latin squares. The independent variables are synchrony between the participant's actions and the visual feedback, i.e., synchronous versus delayed by 500 ms, as well as the predictability of haptic assistance through

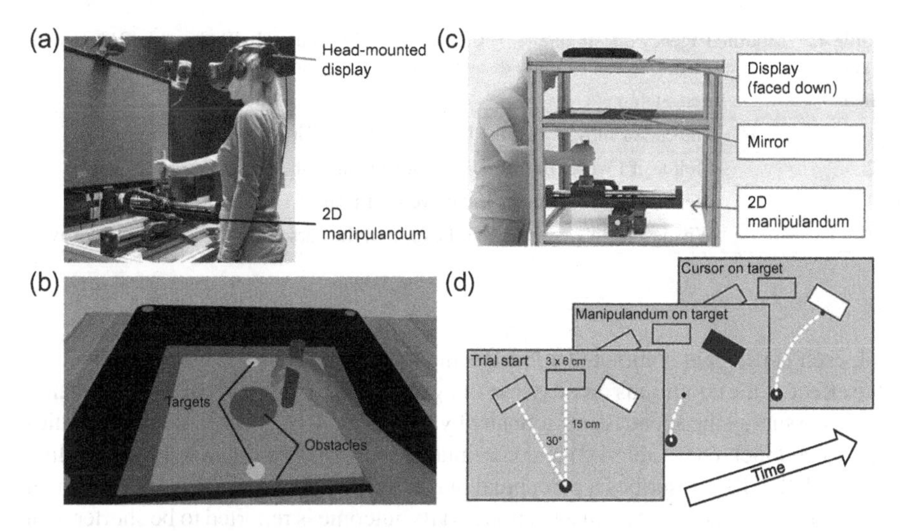

Fig. 4.4 Experimental setup (**a**) and virtual environment (**b**) investigated by Fröhner et al. [20]. Experimental setup (**c**) and virtual environment with decision-making task (**d**) examined by Endo et al. [21]. *Based on figures from* Endo et al. [21] *and* Fröhner et al. [20] (Fröhner et al.: *©2019 IFAC (International Federation of Automatic Control). Reproduced with permission from original publication in IFAC-PapersOnline, 51,34* (https://doi.org/10.1016/j.ifacol.2019.01.036))

the controller, i.e., predictable versus unpredictable. During interaction, an assistive controller reshaped the inertial and damping characteristics of the setup and distributed the task linearly to human and machine, which was randomly altered in the conditions with unpredictable assistance [20]. During a 5-minute interaction period, participants were asked to move the handle from one target to the other avoiding the obstacles (see Fig. 4.4) at a comfortable speed, repetitively. To assess how the different conditions influence user experience, Fröhner et al. [20] acquired subjective data on embodiment with a questionnaire adapted from Longo et al. [16] (similar to Table 4.1) and the control- and sensory-related items from the presence questionnaire of Witmer and Singer [29]. All items were rated on visual analog scales within the virtual environment. Moreover, they measured root mean square jerk, velocity, and compensation forces between user and machine as objective measures. All data were analyzed statistically by calculating ANOVAs.

Endo et al. [21] apply a repeated-measures design manipulating directional correctness of guiding forces, i.e., correct/incorrect/absent, and delayed visual cursor feedback, i.e., 300/500/700 ms, as the independent variables. Participants were asked to perform a decision-making task in which they had to direct a cursor towards one of three equally distant targets (see Fig. 4.4). The mass apparent at the handle was shaped and, depending on the condition, an additional guiding force was either directed to the right target, pseudo-randomly pointing to a wrong target, or absent. The force conditions were randomly presented as blocked trials and target location as well as the non-target force direction were continuously varied for balanced probability. Over-

Table 4.2 Modified agency questionnaire by Caspar et al. [27] as used by Endo et al. [21] rated on a continuous scale ranging from "strongly disagree" to "strongly agree"

Item #	Item statement
1.	The cursor moved just like I wanted it to, as if it were obeying my will
2.	I felt as if I were controlling the movement of the cursor
3.	I felt as if I were causing the movement I saw
4.	Whenever I moved my hand, I expected the cursor to move in the same way

all, each participant completed 405 trials including 135 trials per block. To evaluate the effect of the conditions on the sense of agency, Endo et al. [21] used visual analog scales to survey the agency questionnaire by Caspar et al. [27] at the end of each guiding force block (see Table 4.2) and continuously investigate the intentional binding effect (IBE). IBE describes a perceptual bias related to the sense of agency in which the time experienced between an action and its outcome is reported to be shorter than the real time-lapse if the action was self-triggered by the participant [26]. Considering the perceptual bias as an implicit measure of agency, previous studies showed that IBE can be used to evaluate task sharing between humans and artificial agents [30]. Endo et al. [21] informed participants about the delay between their actions and the outcome prior to the experiment, but stated that this latency would be randomly chosen between 0 and 1000 ms, which were the extrema of the applied visual analog scale. IBE and questionnaire results are analyzed using repeated-measures ANOVAs and relations between both are assessed using Pearson's correlation. To probe the influence of different guiding forces, trial duration, peak interaction force as well as the manipulation share and force [31] are calculated and evaluated by ANOVAs.

4.2.2 Results and Observations

Fröhner et al. [20] found significant main effects of predictability ($F_{1,7} = 6.34$, $p = 0.04$) and synchrony ($F_{1,7} = 43.00$, $p < .001$) on the embodiment questionnaire result. As presented left in Fig. 4.5, embodiment experience was significantly higher predictable (0.76 ± 0.84) than with unpredictable control (0.25 ± 1.13). Stronger embodiment was also experienced in the synchronous conditions (0.96 ± 0.97) compared to the asynchronous conditions (-0.45 ± 1.00). The right plot in Fig. 4.5 outlines that all subscales of embodiment showed similar patterns as the overall score. An interaction effect of the independent variables was not observed. For the control- and sensory-related items of the presence questionnaire, a significant influence of predictability ($p = 0.03$) with higher results for the predictable controller was discovered. Among the objective measures, significant main effects of predictability ($p = 0.05$) and synchrony ($p < .001$) on jerk, which was lower for predictable con-

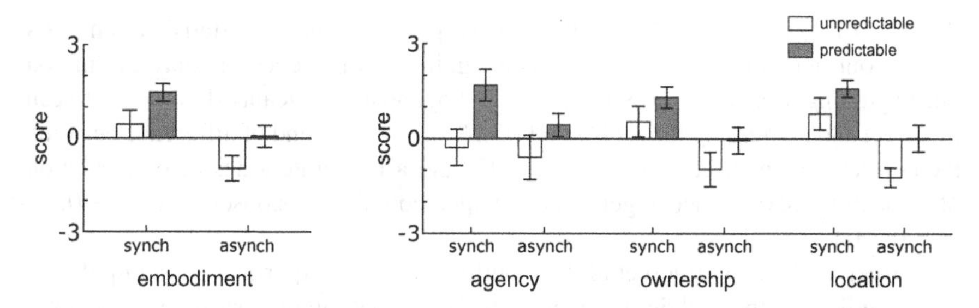

Fig. 4.5 Mean values and standard errors of subjectively experience embodiment (left) and questionnaire subscale ratings (right). *Based on figures from* Fröhner et al. [20] (*©2019 IFAC (International Federation of Automatic Control). Reproduced with permission from original publication in IFAC-PapersOnline, 51,34* (https://doi.org/10.1016/j.ifacol.2019.01.036))

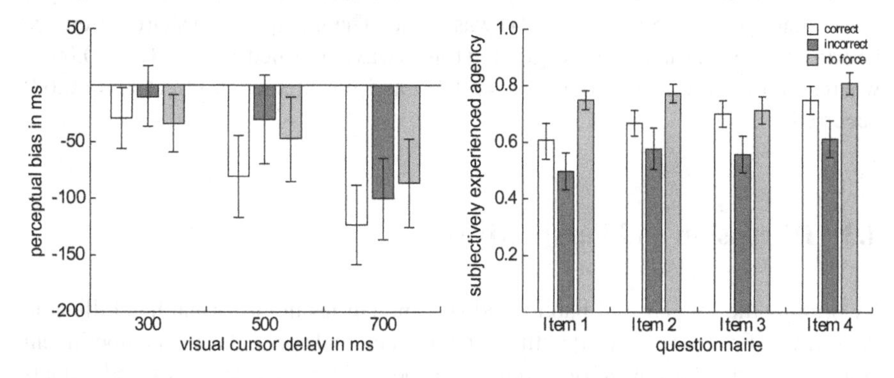

Fig. 4.6 Perceptual bias (left) and subjectively experienced sense of agency (right). *Based on a figure from* Endo et al. [21]

trol, were observed. The controller did not significantly influence compensation force and velocity.

Endo et al. [21] report that the overall perceived delay has been slightly shorter than the actual delay (-60.2 ± 88.9 ms, left in Fig. 4.6) with the tendency being stronger with increasing visual delays. Statistical analysis of IBE data confirmed an effect of the delay in the biased time perception ($F_{2,42} = 11.12$, $p < 0.0005$) with significant differences between 300 and 700 ms ($p < 0.01$) as well as 500 and 700 ms ($p < 0.001$). Regarding guiding force, no main effect was observed ($p = 0.115$), but it showed in interaction with delay ($F_{4,84} = 2.71$, $p < 0.05$): for correctly aligned or absent guiding forces, the time-lapse was perceived as shorter (-77.6 ± 105.8 ms and -56.0 ± 94.9 ms, respectively) compared to incorrectly directed guiding force -46.9 ± 90.8 ms. Endo et al. [21] found evidence for this IBE-like observation regarding estimating 300 and 500 ms delays.

The results regarding subjectively experienced agency are shown right in Fig. 4.6: a self-reported sense of agency was lower with the incorrect guiding force (0.56 ± 0.26) than with the correct guiding force (0.68 ± 0.19) and with no guiding force

(0.76 ± 0.13), which is in line with the perceptual bias results. Statistical analysis on the four questionnaire items reveals a significant difference in delay estimation due to guiding force ($F_{2,42} = 9.62$, $p < 0.0005$) and a significant difference between incorrect and no guiding force ($p < 0.005$). Moreover, a trend of difference between the correct and incorrect forces ($p < 0.07$) and a moderately negative correlation of subjectively experienced agency and the perceptual bias is observed ($r = -0.41$, $p < 0.001$).

Further analysis by Endo et al. [21] outlined that the experimental manipulation did not influence the trial duration ($p > 0.24$) and that interaction force was neither influenced by guiding force ($p = 0.31$) nor by delay ($p = 0.10$). With correct guiding force, participants were responsible for manipulating the cursor by $77.4 \pm 6.2\%$, while their contribution was higher with incorrect guiding force ($80.3 \pm 4.6\%$) due to the lacking assistance. Accordingly, the main effect of guiding force on manipulation share ($F_{1,21} = 5.06$, $p < 0.04$) was found. The manipulation share difference between the correct and incorrect guiding forces was confirmed post hoc ($p < 0.001$), whereas no main effect of delay ($p = 0.33$) and no interaction effect ($p = 0.68$) occurred.

4.3 Discussion and Perspectives

By probing the influence of haptic feedback modalities in the virtual hand illusion, the study by Fröhner et al. [13] outlines the fundamental applicability of embodiment as an engineering design metric. Even though bodily illusions vary across individuals [32], the questionnaire results show a significantly stronger embodiment experience if haptic feedback, i.e., vibrotactile or force feedback, is provided. Since there is no significant difference between the effects of vibrotactile and force feedback, the subsequent engineering decision might be to go for simpler and cheaper vibrotactile feedback. Yet, such decisions should always include an extensive analysis of the given interaction scenario: discrete vibrotactile feedback might suffice for pick-and-place tasks as investigated by Fröhner et al. [13], but maybe not for more ecologically valid situations requiring higher mechanical fidelity [3]. The proprioceptive drift results were close to, but not yet statistically significant, which could be due to a predominance of vision [33, 34] and advocates for a critical reconsideration of proprioceptive drift measurement as motivated by Christ and Reiner [35].

Extending the line of thought, Fröhner et al. [20] demonstrate the potential of embodiment as a quality metric of human-in-the-loop control. Subjective embodiment measures and subscales of presence are found to be significantly influenced by the predictability of control. Higher controllability due to predictable behavior seems to support the interaction of human and machine, which is underlined by the observation of lower jerk levels. Witmer and Singer [29] report a positive relation between presence and task performance, which is similarly found by Fröhner et al. [20] as predictability has a significant influence on the control- and sensory-related factors of presence, reduces jerk, and increases velocity. While results are generally in line with previous works [16, 27], the study agrees with the critical perspective of Fröhner et al. [13] that further research is required to improve our understanding of how the subjective metrics are influenced by the technical solutions. Moreover, the study raised interest with respect to the sense of agency since the agency subscale of the questionnaire was influenced by predictability less distinctly than expected.

For a deeper analysis of the influence of haptic assistance on the sense of agency, Endo et al. [21] applied an adapted intentional binding effect paradigm. Their results show that assistance is perceived as collaborative if it matches the objectives of the user. This suggests the sense of agency as a promising metric of quality and experience in physical human-machine interaction. This conclusion and the observed intentional binding effect are in line with previous research on collective agency [28] and human-in-the-loop experiments [27]. Besides the perceptual bias, subjectively experienced agency and reduced forces demanded from the user underline the supportiveness of the correctly aligned assistive force.

The three studies considered in this chapter outline that haptic feedback modality, control predictability, correct assistance, and system delay are important modulators of embodiment and agency. While embodiment might represent a more general design metric, the sense of agency could help to objectively evaluate assistive control and provide users with intuitive support, i.e., without noticing the actions of the other agent. Future studies should involve further potential modulators as well as ecologically valid tasks and scenarios to advance the user-centric capabilities of experimental paradigms. To this end, virtual reality paradigms open up a wide range of possible implementations and experimental manipulations, which would not be possible using real hardware.

References

1. Sallnäs, E.L., Rassmus-Gröhn, K., Sjöström, C.: Supporting presence in collaborative environments by haptic force feedback. ACM Trans. Comput. Hum. Interact. **7**(4), 461–476 (2000). https://doi.org/10.1145/365058.365086
2. Sadowski, W., Stanney, K.: Presence in virtual environments (2002)
3. Nostadt, N., Abbink, D.A., Christ, O., Beckerle, P.: Embodiment, presence, and their intersections: teleoperation and beyond (submitted). ACM Trans. Hum. Robot Interact. (2020)
4. Basdogan, C., Ho, C.H., Srinivasan, M.A., Slater, M.: An experimental study on the role of touch in shared virtual environments. ACM Trans. Comput. Hum. Interact. (TOCHI) **7**(4), 443–460 (2000)
5. Stone, R.: Haptic feedback: a brief history from telepresence to virtual reality. In: Haptic Human-Computer Interaction, pp. 1–16 (2001)
6. Pacchierotti, C., Sinclair, S., Solazzi, M., Frisoli, A., Hayward, V., Prattichizzo, D.: Wearable haptic systems for the fingertip and the hand: taxonomy, review, and perspectives. IEEE Trans. Haptics **10**(4), 580–600 (2017)
7. Choi, W., Li, L., Satoh, S., Hachimura, K.: Multisensory integration in the virtual hand illusion with active movement. BioMed Res. Int. **2016** (2016)
8. Hara, M., Nabae, H., Yamamoto, A., Higuchi, T.: A novel rubber hand illusion paradigm allowing active self-touch with variable force feedback controlled by a haptic device. IEEE Trans. Hum. Mach. Syst. **46**(1), 78–87 (2016)

9. Sanchez-Vives, M.V., Slater, M.: From presence to consciousness through virtual reality. Nat. Rev. Neurosci. **6**(4), 332–339 (2005)
10. Padilla-Castaeda, M.A., Frisoli, A., Pabon, S., Bergamasco, M.: The modulation of ownership and agency in the virtual hand illusion under visuotactile and visuomotor sensory feedback. Presence: Teleoperators Virtual Environ. **23**(2), 209–225 (2014)
11. Rognini, G., Blanke, O.: Cognetics: robotic interfaces for the conscious mind. Trends Cogn. Sci. **20**(3), 162–164 (2016)
12. Beckerle, P., Castellini, C., Lenggenhager, B.: Robotic interfaces for cognitive psychology and embodiment research: a research roadmap. Wiley Interdiscip. Rev. Cogn. Sci. **10**(2), e1486 (2019)
13. Fröhner, J., Salvietti, G., Beckerle, P., Prattichizzo, D.: Can wearable haptic devices foster the embodiment of virtual limbs? IEEE Trans. Haptics (2018)
14. Prattichizzo, D., Chinello, F., Pacchierotti, C., Malvezzi, M.: Towards wearability in fingertip haptics: a 3-DoF wearable device for cutaneous force feedback. IEEE Trans. Haptics **6**(4), 506–516 (2013)
15. Ismail, M.A.F., Shimada, S.: Robot and illusion under delayed visual feedback: relationship between the senses of ownership and agency. PLOS ONE **11**(7), e0159,619 (2016)
16. Longo, M.R., Schüür, F., Kammers, M.P.M., Tsakiris, M., Haggard, P.: What is embodiment? A psychometric approach. Cognition **107**, 978–998 (2008)
17. Shimada, S., Fukuda, K., Hiraki, K.: Rubber hand illusion under delayed visual feedback. PLoS ONE **4**(7) (2009)
18. Kass, R.E., Raftery, A.E.: Bayes factors. J. Am. Stat. Assoc. **90**(430), 773–795 (1995)
19. Huynh, T.V., Bekrater-Bodmann, R., Fröhner, J., Vogt, J., Beckerle, P.: Robotic hand illusion with tactile feedback: unravelling the relative contribution of visuotactile and visuomotor input to the representation of body parts in space. PloS One **14**(1), e0210,058 (2019)
20. Fröhner, J., Beckerle, P., Endo, S., Hirche, S.: An embodiment paradigm in evaluation of human-in-the-loop control. IFAC-PapersOnLine **51**(34), 104–109 (2019)
21. Endo, S., Fröhner, J., Music, S., Hirche, S., Beckerle, P.: Effect of external force on agency in physical human-machine interaction. Front. Hum. Neurosci. **14** (2020)
22. Hamaya, M., Matsubara, T., Noda, T., Teramae, T., Morimoto, J.: Learning assistive strategies for exoskeleton robots from user-robot physical interaction. Pattern Recognit. Lett. **99**, 67–76 (2017)
23. Erdogan, A., Argall, B.D.: The effect of robotic wheelchair control paradigm and interface on user performance, effort and preference: an experimental assessment. Robot. Auton. Syst. **94**, 282–297 (2017)
24. Limerick, H., Coyle, D., Moore, J.W.: The experience of agency in human-computer interactions: a review. Front. Hum. Neurosci. **8**, 643 (2014)
25. Beckerle, P., Salvietti, G., Unal, R., Prattichizzo, D., Rossi, S., Castellini, C., Hirche, S., Endo, S., Ben Amor, H., Ciocarlie, M., Mastrogiovanni, F., Argall, B.D., Bianchi, M.: A human-robot interaction perspective on assistive and rehabilitation robotics. Front. Neurorobotics **11**(24) (2017)
26. Haggard, P., Clark, S., Kalogeras, J.: Voluntary action and conscious awareness. Nat. Neurosci. **5**(4), 382–385 (2002)
27. Caspar, E.A., Cleeremans, A., Haggard, P.: The relationship between human agency and embodiment. Conscious. Cogn. **33**, 226–236 (2015)
28. Dewey, J.A., Knoblich, G.: Do implicit and explicit measures of the sense of agency measure the same thing? PloS One **9**(10), e110,118 (2014)
29. Witmer, B.G., Singer, M.J.: Measuring presence in virtual environments: a presence questionnaire. Presence **7**(3), 225–240 (1998)
30. Berberian, B., Sarrazin, J.C., Le Blaye, P., Haggard, P.: Automation technology and sense of control: a window on human agency. PLoS One **7**(3), e34,075 (2012)
31. Donner, P., Endo, S., Buss, M.: Physically plausible wrench decomposition for multieffector object manipulation. IEEE Trans. Robot. **34**(4), 1053–1067 (2018)

32. Marotta, A., Tinazzi, M., Cavedini, C., Zampini, M., Fiorio, M.: Individual differences in the rubber hand illusion are related to sensory suggestibility. PloS One **11**(12), e0168,489 (2016)
33. Botvinick, M., Cohen, J.: Rubber hands 'feel' touch that eyes see. Nature **391**, 756 (1998)
34. Hecht, D., Reiner, M.: Sensory dominance in combinations of audio, visual and haptic stimuli. Exp. Brain Res. **193**(2), 307–314 (2009)
35. Christ, O., Reiner, M.: Perspectives and possible applications of the rubber hand and virtual hand illusion in non-invasive rehabilitation: technological improvements and their consequences. Neurosci. Biobehav. Rev. **44**, 33–44 (2014)

Part III
Lower Limbs

Chapter 5
Robotic Leg Experience

Abstract While a large body of research investigates upper limb bodily experience, knowledge concerning the lower limbs is similarly important and can benefit from human-in-the-loop experiments. This chapter inquires specific requirements, presents an appropriate system design, and confirms the transferability of robotic limb illusions to the lower limbs. An evaluation study shows the occurrence of a robotic leg illusion based on subjective embodiment as well as through measuring the proprioceptive drift. The study substantiates the importance of motion synchronization and that robot-related acoustical stimulation is possibly relevant in terms of multisensory integration and embodiment experience. Interrelations of these factors and the design and control of the human-in-the-loop experiment are discussed and alternative approaches are outlined.

5.1 Lower Limb Body Experience

While the rubber hand illusion [1] is established as an experimental paradigm to explore embodiment [2], the lower limbs receive much less coverage in research, which becomes obvious from many reviews [2, 3]. However, translating the paradigm to the lower limbs is possible as has been shown by studies on rubber foot illusions (RFI) [4] that observed similar multisensory integration processes [5, 6]. In a comparison study, Flögel et al. [7] conclude that there are neither qualitative nor quantitative differences between both types of illusions.

Approaching the lack of lower limb embodiment studies, Beckerle et al. [8, 9] extended the concept of the paradigm to the whole leg considering motion and control through robotic augmentation. Through this, they aimed to analyze embodiment triggering the sense of agency in consequence of the actual movement, which was shown to be an important factor for altered upper limb bodily awareness [10–12]. Going beyond classical designs [2], the resulting setup, i.e., the RobLI, was used by Penner et al. to conduct an interactive human-in-the-loop experiment [13]. It enabled variation of motion synchrony and was prepared to consider distance to the robotic limb and vibrotactile feedback for analyses of multisensory integration [9, 13].

P. Beckerle, *Human-Robot Body Experience*, Springer Series on Touch and Haptic Systems, https://doi.org/10.1007/978-3-030-38688-7_5

The result of these efforts was the very first examination of the RobLI by Penner et al. [13], which is presented subsequently as an example of lower limb human-in-the-loop experiments. To enable this pioneering step, the mechatronic design was improved to mitigate issues like system-intrinsic delays due to constrained robot dynamics and human motion tracking as presented in Sect. 5.2. The RobLI experiment described in Sect. 5.3 analyzed whether embodiment is similarly modulated as in RHI and RobHI. In the upper limbs, the illusion rather rapidly declines if robot hand motions are delayed more than 500 ms compared to the real hand [14] or if the distance between the limbs goes beyond 0.3 m [15–18]. Since these are only the first steps towards understanding lower limb bodily illusions, cognitive models are suggested subsequently in Chap. 6 for a deeper analysis.

5.2 Implementation of Robotic Leg Illusions (RobLI)

Considering the design and control concepts proposed by Beckerle et al. [8, 9], Penner et al. presented an elaborated mechatronic RobLI design. This section gives a global description of this latest version of the RobLI setup. Detailed information and parameters are found in the work of Penner et al. [13].

Mechanically, a double-inverted pendulum realizes the robotic leg with the upper and lower links representing shank and thigh, respectively. Robot joint motions imitate participant motions measured with an instrumented knee bandage. The system is shown in Fig. 5.1 and covered with hull of a shop-window mannequin and the same trousers as the participant to meet the requirement of similar outer appearance [8, 9, 19].

5.2.1 Mechatronic Hardware

To overcome limitations of the previous designs [8, 9] in meeting temporal and spatial requirements [19], Penner et al. revised the implementation of the knee joint [13]. Applying remote actuation for all joints, a more powerful DC motor could be used for the knee and hidden under the surface of the experimental setup as the ankle actuator. Figure 5.1a, b shows how the knee joint is actuated via Bowden cables and the ankle joint is driven through a timing belt. Both DC motors were combined with appropriate gearboxes, driven by four-quadrant DC servo amplifiers, and equipped with incremental encoders measuring motor positions. Additionally, an encoder is applied to the knee joint to tackle potential deviations introduced due to Bowden cable elasticity.

The instrumented bandage measuring participant motions is implemented by two inertial measurement units sewed to the thigh and shank parts of a commercially available knee bandage as shown in Fig. 5.1d. Fusing the acquired accelerometer and gyroscope data, knee and ankle joint positions are determined and used as desired

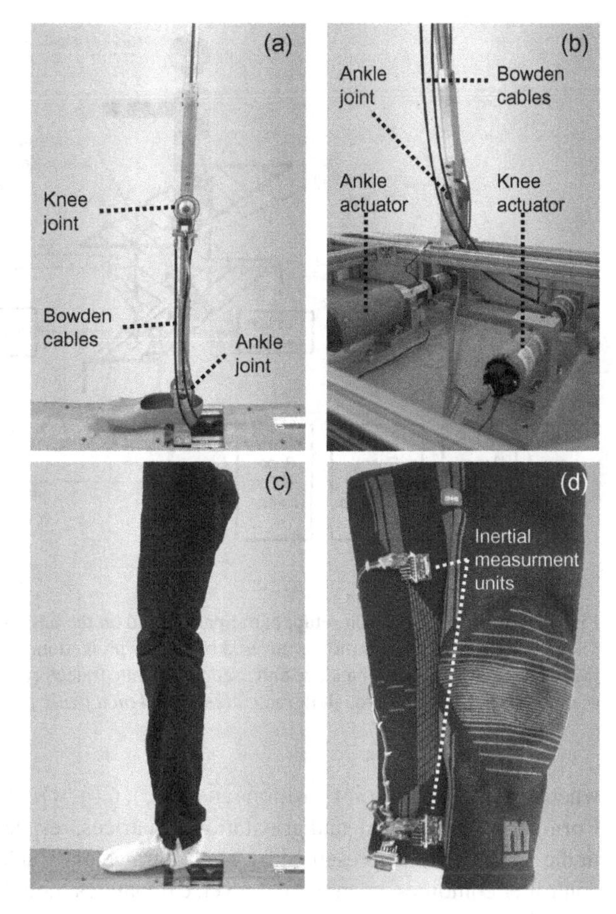

Fig. 5.1 Robotic leg with knee (**a**) and ankle (**b**) joint implementation and actuation as well as with cladding and clothes (**c**). Instrumented bandage (**d**). *Based on figures from* Penner et al. [13]

values for robot motion control. Data acquisition, signal processing, and motion control are implemented on a real-time control platform and operated by a host computer.

5.2.2 Motion Control

As shown in Fig. 5.2, the robotic leg is motion controlled based on sensory data from human motion acquisition and interoceptive sensing in the robot. Robot motion trajectories, i.e., desired knee and ankle positions, are generated by fusing the orientations of the human shank and thigh by a complementary filter [20]. To mirror these desired motions, a computed torque controller is implemented and friction is feedback compensated [13]. Considering friction effects, the structure of the dynamics model of the robotic leg is

$$\tau = M(q)\ddot{q} + C(\dot{q}, q) + G(q) + \tau_f, \tag{5.1}$$

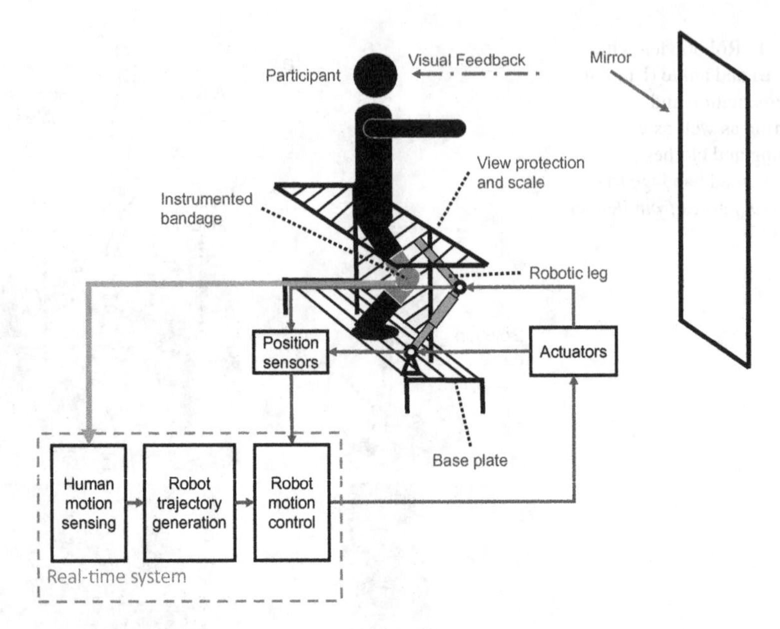

Fig. 5.2 Robotic leg illusion setup: participants stand on the base plate while looking at a mirror. The real leg under consideration is covered by a view protection. Human motions are acquired by IMUs on the instrumented bandage and used to generate trajectories. Measured robot positions are used for motion control through the actuators. *Based on a figure from* Penner et al. [13]

where q describes the joint positions and $M(q)$, $C(\dot{q}, q)$, and $G(q)$ denote the inertial, Coriolis and centrifugal, and gravitational matrices, respectively. The friction torque in the drive train is represented by τ_f. For trajectory tracking, the computed torque controller combines an analytical inverse dynamics model with a PD-type feedback controller:

$$\tau_c = M(q)\left[\ddot{q}_d + k_p\tilde{q} + k_v\dot{\tilde{q}}\right] + C(\dot{q}, q) + G(q) + \tau_{fc}\,. \tag{5.2}$$

To achieve critically damped system dynamics, the control parameter matrices $k_p = diag(k_{p,a} \quad k_{p,k})$ and $k_v = diag(k_{v,a} \quad k_{v,k})$ are set to $k_p = diag(\omega^2 \quad \omega^2)$ and $k_v = diag(2\omega \quad 2\omega)$. In this, ω is the desired exponential decay rate of the tracking error $\tilde{q} = q_d - q$ [21]. The final term τ_{fc} of the control law (5.2) implements compensation of Coulomb and viscous friction via an algorithm by Rinderknecht and Strah [22]. It was adjusted to the RobLI setup by Beckerle et al. [9] and continuously compensates for Coulomb friction and considers viscous friction in case of increasing motion velocities using fuzzy logic.

Aiming at human-in-the-loop experiments, fast and smooth trajectory tracking is crucial, but limited through disturbances occurring during human motion acquisition and motion control, especially by temporal delays. Since such effects constrained the capabilities of earlier versions of the RobLI setup [9, 20, 23], human-machine control performance was examined by Penner et al. [13] measuring the response times of the system during squat motions. The experiment yielded good tracking results and very good motion synchronization. The study also observed system-intrinsic delays, which were rather high for motion initiation (ankle: 393.2 ± 22.6 ms, knee: 369.9 ± 18.7 ms), but appropriate regarding motion completion (ankle: 135.0 ± 14.3 ms, knee: 136.6 ± 25.0 ms).

5.3 Human-in-the-loop Evaluation of the RobLI

Despite remaining technical constraints, Penner et al. demonstrated that illusionary effects can be elicited with the RobLI setup, which enables it for human-in-the-loop experiments [13]. Subsequently, their experimental methods and results concerning their hypotheses:

> Hypothesis 1: If the robotic leg moves synchronous, the proprioceptive drift (PPD) is significantly larger than with asynchronous movement.
>
> Hypothesis 2: If the robotic leg moves synchronous, the subjectively perceived embodiment is significantly stronger than with asynchronous movement.

are described briefly to outline challenges and possibilities.

5.3.1 Materials and Methods

The study relies on data of 31 participants (51.6% female, 48.4% male; average age of 28.3 years, standard deviation of 8.6 years) recruited via e-mail and blackboard notes. All participants expressed that they were naive to the purpose of the experiment and provided written, informed consent. The procedure of the study is in accordance with the Helsinki declaration and was conducted considering recommendations of the ethics commission of Technical University Darmstadt, Germany.

After giving informed consent and a brief training session with instructions of the experimenter, the participants' locations on the base plate were aligned with the measuring scale starting at the inner side of the robotic foot. During all three conditions, the participants observed themselves in the mirror while conducting three times 15 knee bends separated by 15 seconds of break in an upright position. Each condition had a duration of 3.5 min and breaks of five minutes were inserted to ensure recreation and extinguishing the illusion completely.

To cope with interindividual differences of body experience, a repeated-measure experimental design was chosen to compare two conditions: synchronous (synch-near) and asynchronous (asynch-near, delayed by 500 ms) control of the robotic leg. Sequence effects were controlled by Latin squares. The influence of leg locations

Table 5.1 RobLI-adjusted version of the embodiment questionnaire by Longo et al. [24] in English and the version translated to German as used by Penner et al. [13]

Item #	Original statement During the block...	Translated item Während der Durchführung...
1.	... it seemed like I was looking directly at my own leg, rather than at a robotic leg	... schien es, als schaute ich eher auf mein eigenes statt auf ein Roboterbein
2.	... it seemed like the robotic leg began to resemble my real leg	... schien es, als würde das Roboterbein beginnen, meinem echten zu ähneln
3.	... it seemed like the robotic leg belonged to me	... schien es, als gehöre das Roboterbein zu mir
4.	... it seemed like the robotic leg was my leg	... schien es, als sei das Roboterbein mein eigenes
5.	... it seemed like the robotic leg was part of my body	... schien es, als sei das Roboterbein Teil meines Körpers
6.	... it seemed like my leg was in the location where the robotic leg was	... schien es, als befand sich mein eigenes Bein dort, wo das Roboterbein war
7.	... it seemed like the robotic leg was in the location where my leg was	... schien es, als befand sich das Roboterbein dort, wo mein echtes Bein war
8.	... it seemed like a touch on the robotic leg I would perceive as a touch on my own leg	... schien es, als könnte ich Berührungen am Roboterbein als Berührungen an meinem echten Bein wahrnehmen
9.	... it seemed like I could have moved the robotic leg if I had wanted	... schien es, als könne ich das Roboterbein bewegen, wenn ich wollte
10.	... it seemed like I was in control of the robotic leg	... schien es, als habe ich die Kontrolle über das Roboterbein

was considered by adding a condition with increased distance between human and robotic leg (synch-far) motivated by [16]. In this case, the distance between both legs was increased from 17.5 to 35 cm. As bodily illusions are subject to multisensory integration, questionnaire items asking for acoustic influences were complemented.

Considering hypothesis 1, the illusion is assessed objectively using the proprioceptive drift, i.e., the difference of the actual and experienced leg positions. To this end, participants closed their eyes and indicated the position of their right leg with their index finger. PPD was acquired before the experiment (pre-measurement) and after each condition (post-measurement). To assess the subjective experience of embodiment, participants completed a RobLI-adjusted version of the survey by Longo et al. [24]. On a seven-point Likert scale ranging from "1 = Strongly disagree" to "7 = Strongly agree", they reported their body experience regarding ownership, location, and agency regarding the robotic leg as shown in Table 5.1. Results were evaluated by one-way analyses of variances (ANOVA).

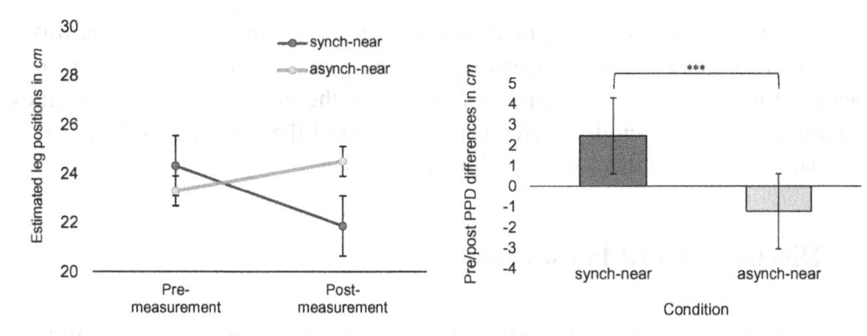

Fig. 5.3 Proprioceptive drift towards the robotic leg: mean estimated real leg positions before (pre-measurement) and after (post-measurement) inducing the illusion (left) and mean drift differences (right). Error bars represent standard errors. *Based on a figure from* Penner et al. [13]

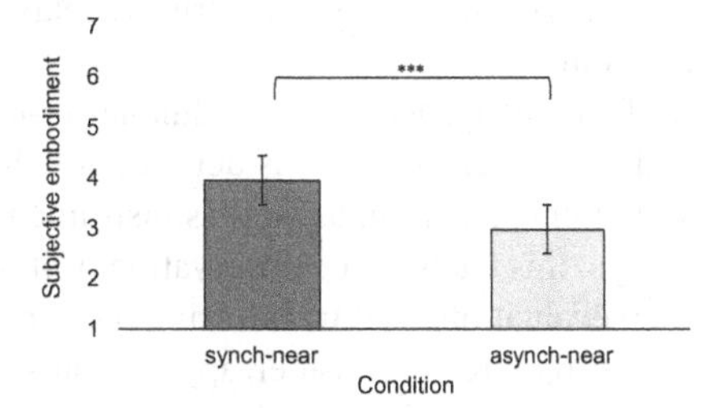

Fig. 5.4 Means (bars) and standard deviations (whiskers) of subjective embodiment experience with respect to the robotic leg.*Based on a figure from* Penner et al. [13]

5.3.2 Results and Observations

Hypothesis 1 expects a synchronous movement (synch-near) to cause a stronger illusionary effect than an asynchronous movement (asynch-near). This is significantly underlined by a paired-samples t-test ($t(30) = 5.45$, $p < .001$). Taking a look at proprioceptive drift data presented left in Fig. 5.3 shows a shift of perceived leg location towards the robotic leg from pre- ($M_{synch-near} = 24.31$, $SD_{synch-near} = 3.23$) to post-measurement results ($M_{synch-near} = 21.86$, $SD_{synch-near} = 4.36$).

Comparing pre- ($M_{asynch-near,pre} = 23.28$, $SD_{asynch-near,pre} = 3.92$) and post-measurement ($M_{asynch-near,post} = 24.50$, $SD_{asynch-near,post} = 3.63$) data in the asynch-near condition points to a drift away from the robotic leg. The difference of post-measurement and pre-measurement data is shown on the right of Fig. 5.3 highlighting the significant difference between the conditions.

Besides PPD changes, subjectively perceived embodiment was expected to be stronger in the synchronous condition (synch-near) according to hypothesis 2. This is supported by a significant difference between the two synchronous conditions ($t(30) = 5.52$, $p < .001$) in the questionnaire items that is outlined in Fig. 5.4.

Beyond probing the two hypotheses, the study further explored factors potentially influencing multisensory integration and the resulting illusionary embodiment. Focus was set on the distance between the human and robotic leg as well as on acoustics. To get insights in the effect of distance, data obtained in the synch-near and synch-far conditions was compared by a paired-samples t-test showing no significant difference neither for PPD nor for subjectively perceived embodiment [13]. With respect to acoustic influences, 29 participants reported in the closing demographic questionnaire that they consciously perceived the sound of the robot leg and half of the participants (15) stated the noise to be disturbing [13].

5.4 Discussion and Perspectives

The study by Penner et al. [13] was the first one transferring the rubber hand illusion (RHI) to the whole leg in general and during motion. This was achieved through an in-depth analysis of human-related requirements based on the insights reported in Sect. 2.3 and a correspondingly improved mechatronic system. An important aspect is the reduction of system-intrinsic delay due to the importance of motion synchronization.

Technically, achieving the human-in-the-loop requirements demands for system- and component-level consideration touching mechanics and actuation, e.g., remote knee joint actuation, as well as instrumentation and control [13]. Despite remaining delays, this leads to the observation of illusionary embodiment in the human-in-the-loop evaluation. Yet, transferring these results to real applications, e.g., prosthetics, would require to consider application-specific human-machine interfaces such as electromyographic control or haptic feedback [2, 25].

From a psychological perspective, it is noteworthy that Penner et al. found synchronous movement (synch-near) to elicit an illusion considering subjective and objective measures, which provides significant evidence. Similar to previous study, motion synchrony was found to be important for the embodiment of foreign objects [5, 7, 18, 24].

While appropriate requirement analysis and corresponding design decisions facilitate human-in-the-loop experiments, further studies are required to understand the (technical) factors modulating embodiment. As shown by Penner et al., further sensory information such as auditory cues might be relevant suggesting the examination of visual-tactile and audio-tactile integration in future studies [13]. Remarkably, human-in-the-loop techniques can help to understand those relations and, in return, be improved themselves based on the results. Moreover, the effect of distance is very interesting for wearable robotics applications and would be worth deeper investigation, e.g., considering the effect of motions on spatial body experience [26] and creation of spatial bodily awareness [27].

References

1. Botvinick, M., Cohen, J.: Rubber hands 'feel' touch that eyes see. Nature **391**, 756 (1998)
2. Beckerle, P., Castellini, C., Lenggenhager, B.: Robotic interfaces for cognitive psychology and embodiment research: a research roadmap. Wiley Interdiscip. Rev.: Cogn. Sci. **10**(2), e1486 (2019)

3. Christ, O., Reiner, M.: Perspectives and possible applications of the rubber hand and virtual hand illusion in non-invasive rehabilitation: technological improvements and their consequences. Neurosci. Biobehav. Rev. **44**, 33–44 (2014)
4. Christ, O., Elger, A., Schneider, K., Beckerle, P., Vogt, J., Rinderknecht, S.: Identification of haptic paths with different resolution and their effect on body scheme illusion in lower limbs. Technically Assisted Rehabilitation (2013)
5. Lenggenhager, B., Hilti, L., Brugger, P.: Disturbed body integrity and the rubber foot illusion. Neuropsychology **29**(2), 205 (2015)
6. Crea, S., D'Alonzo, M., Vitiello, N., Cipriani, C.: The rubber foot illusion. J. NeuroEng. Rehabil. **12**, 77 (2015)
7. Flögel, M., Kalveram, K.T., Christ, O., Vogt, J.: Application of the rubber hand illusion paradigm: comparison between upper and lower limbs. Psychol. Res. **80**(2), 298–306 (2015)
8. Beckerle, P., Christ, O., Wojtusch, J., Schuy, J., Wolff, K., Rinderknecht, S., Vogt, J., von Stryk, O.: Design and control of a robot for the assessment of psychological factors in prosthetic development. In: IEEE International Conference on Systems, Man and Cybernetics (2012)
9. Beckerle, P., Schültje, F., Wojtusch, J., Christ, O.: Implementation, Control and User-Feedback of the Int^{2}Bot for the Investigation of Lower Limb Body Schema Integration. In: IEEE International Symposium on Robot and Human Interactive Communication (2014)
10. Tsakiris, M., Prabhu, G., Haggard, P.: Having a body versus moving your body: how agency structures body-ownership. Conscious. Cogn. **15**(2), 423–432 (2006)
11. Caspar, E.A., de Beir, A., Magalhães Da Saldanha da Gama, P.A., Yernaux, F., Cleeremans, A., Vanderborght, B.: New frontiers in the rubber hand experiment: when a robotic hand becomes one's own. Behav. Res. Methods **47**(3), 744–755 (2015)
12. Caspar, E.A., Cleeremans, A., Haggard, P.: The relationship between human agency and embodiment. Conscious. Cogn. **33**, 226–236 (2015)
13. Penner, D., Abrams, A.M.H., Overath, P., Vogt, J., Beckerle, P.: Robotic leg illusion: system design and human-in-the-loop evaluation. IEEE Trans. Hum. Mach. Syst. (2019)
14. Rohde, M., Di Luca, M., Ernst, M.O.: The rubber hand illusion: feeling of ownership and proprioceptive drift do not go hand in hand. PLoS ONE **6**(6) (2011)
15. Holle, H., McLatchie, N., Maurer, S., Ward, J.: Proprioceptive drift without illusions of ownership for rotated hands in the rubber hand illusion paradigm. Cogn. Neurosci. **2**(3–4), 171–178 (2011)
16. Lloyd, D.M.: Spatial limits on referred touch to an alien limb may reflect boundaries of visuo-tactile peripersonal space surrounding the hand. Brain Cogn. **64**, 104–109 (2007)
17. Shimada, S., Fukuda, K., Hiraki, K.: Rubber hand illusion under delayed visual feedback. PLoS ONE **4**(7) (2009)
18. Tsakiris, M., Haggard, P.: The rubber hand illusion revisited: visuotactile integration and self-attribution. J. Exp. Psychol.: Hum. Percept. Perform. **31**(1), 80–91 (2005)
19. Beckerle, P., De Beir, A., Schürmann, T., Caspar, E.A.: Human body schema exploration: analyzing design requirements of robotic hand and leg illusions. In: IEEE International Symposium on Robot and Human Interactive Communication (2016)
20. Schürmann, T., Overath, P., Christ, O., Vogt, J., Beckerle, P.: Exploration of lower limb body schema integration with respect to body-proximal robotics. In: IEEE International Forum on Research and Technologies for Society and Industry Leveraging a Better Tomorrow (2015)
21. Kelly, R., Santibáñez Davila, V., Loría Perez, J.A.: Control of Robot Manipulators in Joint Space. Springer (2005)
22. Rinderknecht, S., Strah, B.: Simple and effective friction compensation on wheeled inverted pendulum systems. In: Euro-Mediterranean Conference on Structural Dynamics and Vibroacoustics (2013)
23. Schültje, F., Beckerle, P., Grimmer, M., Wojtusch, J., Rinderknecht, S.: Comparison of trajectory generation methods for a human-robot interface based on motion tracking in the Int^{2}Bot. In: IEEE International Symposium on Robot and Human Interactive Communication (2014)
24. Longo, M.R., Schüür, F., Kammers, M.P.M., Tsakiris, M., Haggard, P.: What is embodiment? A psychometric approach. Cognition **107**, 978–998 (2008)

25. Castellini, C., Artemiadis, P.K., Wininger, M., Ajoudani, A., Alimusaj, M., Bicchi, A., Caputo, B., Craelius, W., Došen, S., Englehart, K.B., Farina, D., Gijsberts, S., Godfrey, S.B., Hargrove, L.J., Ison, M., Kuiken, T.A., Markovic, M., Pilarski, P.M., Rupp, R., Scheme, E.: Proceedings of the first workshop on peripheral machine interfaces: going beyond traditional surface electromyography. Front. Neurorobot. **5**(22), 1–17 (2014)
26. Iriki, A., Tanaka, M., Iwamura, Y.: Coding of modified body schema during tool use by macaque postcentral neurones. Neuroreport **7**(14), 2325–2330 (1996)
27. Nostadt, N., Abbink, D.A., Christ, O., Beckerle, P.: Embodiment, presence, and their intersections: teleoperation and beyond (submitted). ACM Trans. Hum. Robot. Interact. (2020)

Chapter 6
Cognitive Models of Body Experience

Abstract Besides considering human-robot body experience as a metric in robot and control design, understanding it in a broader sense and context is of psychological and technical interest. This chapter discusses the potential of cognitive models of body experience in robotics. Approaches like Bayesian or connectionist models might enable adapting assistive robots to their users' body experiences or to endowed humanoid robots with human-like body representations. Besides improving interaction capabilities, this might yield human-like action-perception versatility. However, cognitive body experience models do not yet achieve sufficient accuracy, individualization, and online capabilities. As an exemplary approach, a Bayesian cognitive model of crossmodal sensory processing during the rubber foot illusion is presented. Estimations of empirical results with an extended Bayesian causal inference model are improved by involving empirically informed prior information.

6.1 Potentials of Cognitive Body Models in Robotics

Multisensory integration is central to human body experience [1, 2] and assumed to be a Bayesian process [3–6]. Based on the perspective by Schürmann et al. [7], this section discusses the potential of computational accounts to model cognitive processes, in general, and multisensory integration, in particular.

Cognitive models might relate to one or multiple cognitive functions [8] and are often interpreted with respect to Marr's levels of analysis, i.e., the computational, algorithmic, and implementational levels [9]. The computational level describes the purpose of information processing, the algorithmic level focuses on how this is solved, and the implementation level can comprise biological and/or technical implementations. Cognitive models of multisensory integration processes rely on information about sensorimotor precision of the contributing modalities [10] and facilitate estimating integration results from prior knowledge and sensory data.

Schürmann et al. [7] argue that mathematical models and simulations of multisensory integration processes can push frontiers in several areas of robotics, e.g., by endowing (humanoid) robots with more human-like capabilities and improving tight interaction with assistive devices. Regarding assistive devices, multisensory

P. Beckerle, *Human-Robot Body Experience*, Springer Series on Touch and Haptic Systems, https://doi.org/10.1007/978-3-030-38688-7_6

integration processes determine whether bodily illusions occur or not [1, 2, 11], which substantiates their importance to improve the embodiment [2, 12, 13] and provide their users with the sense of agency [14–16]. Endowing humanoids with a human-like body schema is highly interesting from the perspective of controls. Knowing their own bodies could help robots to keep safe distances and reach for targets [17, 18] as well as in hand/tool-eye coordination robots [19]. Psychological studies imply a direct relation between representations of the human body and the reachable environment, i.e., the peripersonal space [20, 21]. The flexible discrimination between the self and the environment including the integration of tools seems rather underdeveloped in contemporary humanoid robots [22, 23].

Motivated by the potential of such capabilities, recent research studies robotic self-perception [19, 24–26] and model learning [27], analyzes interrelations of human body experience and robotics [13, 23, 28], and develops cognitive body experience models, e.g., Bayesian models of bodily illusions [29, 30]. Schürmann et al. [7] discuss Bayesian and connectionist approaches and focus on cognitive models of rubber/robot limb illusions.

6.1.1 Modeling Approaches

Motivated by the assumption that crossmodal information is integrated based on Bayesian principles [10], recently developed cognitive models aim at estimating the proprioceptive drift from empirical data [29, 30]. While these models generate viable estimates, they still show potential for accuracy improvement by model extension [30] and are constrained to offline application on population level [29, 30]. So far, Bayesian models are often interpreting multisensory processing on the computational level [10, 29, 30] although cognitive body experience models might also benefit from extension to deeper modeling levels [7, 31]. Combining computational- and algorithmic-level perspectives in modeling could improve and explain the prediction of seemingly error-prone or paradoxical behavior [32, 33] and thereby inspire and guide technical development [7].

Connectionism provides an alternative approach commonly relying on artificial neural networks, which might even be implemented in a neurally plausible way to create "brain-like" systems by biological or technical hardware [7, 34]. Remarkably, connectionist approaches to multisensory integration [35, 36] show certain similarities with Bayesian models and can be expected to achieve comparable prediction capabilities [7, 34].

With respect to human body experience, Bayesian models are promising due to their relation to human sensorimotor behavior [37, 38] while the relation of connectionist approaches to developmental psychology and robotics [39, 40] is a rationale for applying those. Moreover, both techniques can be applied for fundamental research of human sensorimotor behavior generation and to control artificial systems. Compared to connectionism, Bayesian approaches more frequently relate to human performance, which supports human-like designs [7].

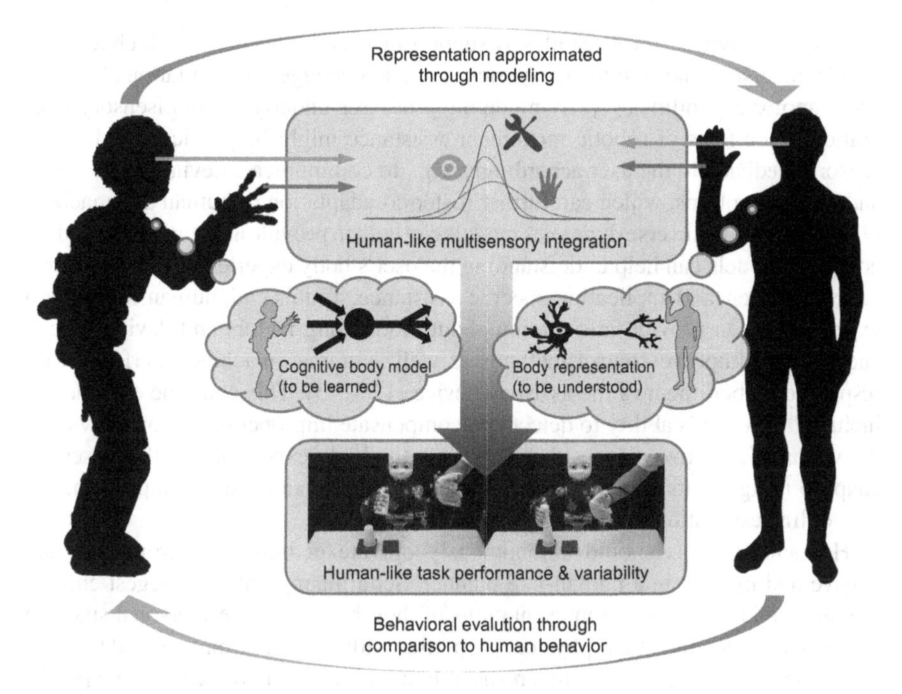

Fig. 6.1 Understanding human body experience and approaching human-like capabilities in robots: multisensory information from human/robot perception is processed by a cognitive body experience model during sensorimotor manipulation. The model can be applied in robot control to generate human-like body representations and behaviors. Iterative research of behavioral data sheds light on fundamental human cognitive processes, which fosters model improvement in turn. *Based on a figure from* Schürmann et al. [7]

6.1.2 Robotic Applications

Schürmann et al. [7] expect cognitive models to improve our understanding of human body experience and to advance robot capabilities. Particularly, they focus on assistive robotic devices and humanoid robots to discuss potentials and challenges. Figure 6.1 outlines an iterative research approach in which both endeavors benefit from each other to, finally, yield human-like multisensory integration and behavior generation in robots. According to Hoffmann et al. [23], knowledge about the own physical self and sensorimotor mapping are required to make robots perform a goal-directed action. While physical body knowledge might be explicitly represented in terms of kinematics and dynamics, sensorimotor mappings and self-recognition are usually implemented implicitly [23]. Explorative behaviors can foster motor and cognitive development based on internal body representations in artificial agents [28], e.g., through MOdular Selection And Identification for Control (MOSAIC) models [41]. Despite these promising steps, implementing and exploiting human-like body representations in robots remains challenging [42].

Assistive devices aim at seamlessly supporting users' movements, which requires a deep understanding of human experience and knowledge representation about the users' motor capabilities [7]. By modeling processes underlying multisensory integration, the effects of robotic movement assistance might be predicted and adjust sensory feedback to the user accordingly, e.g., to communicate deviations of actual and desired motions, which can further foster co-adaptation of human and machine [7, 13, 43]. While inverse dynamics models can help to provide assistance-as-needed, cognitive models can help understanding the user's body experience and consider it to achieve user- and application-specific assistance. To this end, human-in-the-loop experiments can provide information about modulating factors and device experience [43] to improve control methods as well as human-machine interfaces with respect to embodiment of the assistive device [1, 7, 13]. This could be extended to include the device's ability to detect and compensate improper operation. However, the vision of assistive devices that anticipate the body experience of their users to adapt to it highlights that current cognitive models demand distinct improvements, e.g., online estimation of individual experience [7].

Humanoid robots should autonomously operate in human environments and behave and interact in a human-like manner. Schürmann et al. [7] suggest endowing humanoid robots with representations of their bodies and peripersonal space to make them understand their physical properties within the environment and to adapt their interactive behaviors. In this respect, human-like behavior generation appears promising and could be combined with approaches from developmental robotics [7, 40, 42, 44]. Humanoid robots are recently prepared to develop several forms of body representations [45–47] and learn movement generation [48]. Pioneering work indicates robots with whole-body artificial skin can learn peripersonal space models through physical contact with the environment [17, 18], which can be used for online control assuring safety or supporting reaching for objects [17, 18]. To achieve human-like variability in behavior generation and human-robot interaction, established kinematics and dynamics models might be complemented by psychologically motivated cognitive models [7]. Yet, research lacks a formal benchmarking method juxtaposing robot performance and human behavior continuously as shown in Fig. 6.1. For instance, robots could participate the sensorimotor experiment suggested by Körding and Wolpert [49] to explore whether their programming provides human-like behavioral variability and online adaptation.

6.2 Bayesian Modeling of the Rubber Foot Illusion (RFI)

As discussed above, Bayesian cognitive modeling can provide a deeper understanding of the human mind and improve the capabilities of artificial cognitive systems, e.g., humanoid or wearable robots. To outline what can be achieved with state-of-the-art models, this section presents results obtained by Schürmann et al. [30]. The applied modeling approach focuses on crossmodal processing of sensory information in rubber foot illusion experiments. Schürmann et al. [30] proposed two methods to

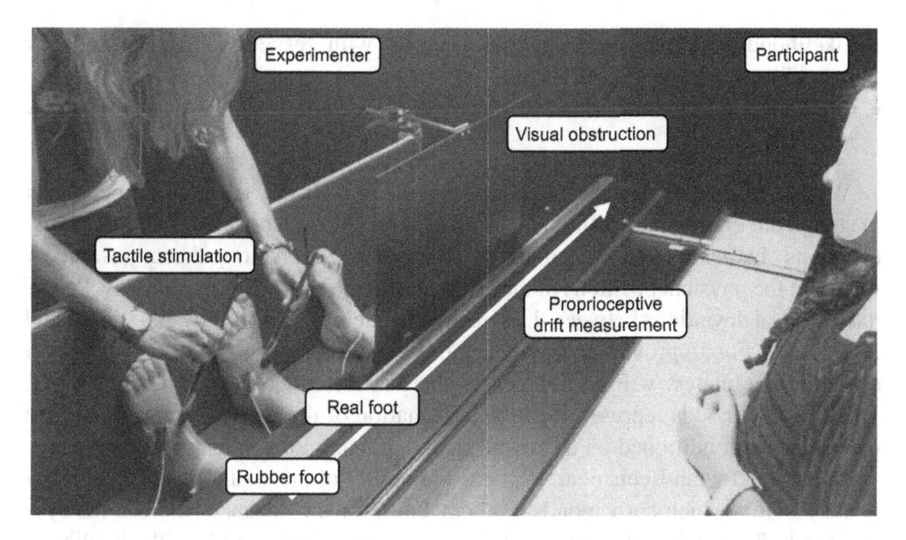

Fig. 6.2 Rubber foot illusion experimental setup implemented by Christ et al. [50]. *Based on a figure from* Schürmann et al. [30]

consider informed prior distributions on limb location and compared them against each other as well as a previously proposed uniform model by Samad et al. [29].

The empirical data that is used to compare model estimates stems from two previous studies by Flögel et al. [51] and Christ et al. [50]. As shown in Fig. 6.2, a horizontally aligned RFI setup was used. To determine the proprioceptive drift, experienced foot positions were measured by verbally instructing the experimenter to stop a light that was moving horizontally above the limbs. Schürmann et al. [30] aimed to predict the empirical results of the study by Flögel et al. [51], which confirmed the existence of the RFI effect, and used prior data from Christ et al. [50] to create an empirically informed variant of the cognitive model.

6.2.1 Cognitive Models of Multisensory Integration

Transferring the model of Samad et al. [29] from the RHI to the RFI, Schürmann et al. [30] considered the increased distance from the eyes to the feet compared to the hands and the different sensory precision. The applied Bayesian causal inference model represents sensory information through probability distributions:

- Prior probabilities of common cause $p(C = 1)$ and separate causes $p(C = 2)$.
- Prior localization distribution $p(X)$.
- Prior visuotactile distribution $p(T)$.
- Likelihood of perceived visual information with respect to localization $p(\chi_V|X)$.
- Likelihood of perceived proprioceptive information $p(\chi_P|X)$.

- Likelihood of perceived visual information with respect to tactile feedback $p(\zeta_V|T)$.
- Likelihood of perceived tactile information with respect to tactile feedback $p(\zeta_T|T)$.

While C is a binary variable indicating whether the sensory inputs are experienced to stem from common or different causes, parameters are considered Gaussian distributions characterized by mean and standard deviation. Among those, χ_V and χ_P represent the physical positions of the artificial and the real limb, respectively. Potential temporal deviations between the sensory channels are denoted by ζ_V (visual) and ζ_T (tactile). For improved comprehensibility, these spatial position and temporal deviation parameters will be denoted as the parameter set $\Xi = \{\chi_V, \chi_P, \zeta_V, \zeta_T\}$. The four likelihoods representing how information of the individual modalities are perceived are conditioned on the actual spatial and temporal values of X and T.

According to van Beers et al. [52], the variability of the visual system is approximately 0.36°, which corresponds to about 1 mm for the distance between the eyes and hands. To consider the increased distance between eyes and feet, the likelihood of perceived visual information $p(\chi_V)$ was modified an expected deviation of 0.6°, which results to a standard deviation of 0.012 m for an average eye-foot distance of 1.125 m and a mean of 0.3 m, i.e., the actual rubber foot position. Temporal estimation standard deviations are set to 20 ms for visual and tactile information referring to Hirsh and Sherrick [53]. Generalizing the RHI parameters of Samad et al. [29], Schürmann et al. [30] set the proprioceptive likelihood distribution $p(\chi_P)$ to a mean of 0.5 m and a standard deviation of 0.015 m. Accounting for the synchronous experimental condition, the temporal distributions regarding vision $p(\zeta_V)$ and touch ζ_T are parameterized with a mean of 0.5 s and a standard deviation of 0.02 s [29, 30]. Since participants were not informed about when tactile stimulation would occur during the experiment, an uninformed visuotactile prior distribution $p(\zeta_V)$ with a mean of 0.5 s and a standard deviation of 10.0 s is used [30]. With a value of 0.5, the prior probability of a common cause was left uninformed as well.

Samad et al. [29] assigned uniform prior distributions across a constrained area to the incoming sensory data Ξ, which would imply equal plausibility for all considered hand locations and visuotactile delays. Schürmann et al. [30] argue that this is not reflecting the actual situation-related knowledge of human participants and hypothesized that the overestimation of proprioceptive drift observed in the results of Samad et al. [29] could be caused by missing prior knowledge. To probe whether an informed prior would improve estimation accuracy, Schürmann et al. [30] consider conceptually informed and empirically informed models in comparison to the original uniform model that does not use prior data. In the conceptually informed model, the real limb location (0.5 m) and the proprioceptive precision (0.015 m) are set as the mean and standard deviation of the Gaussian prior distribution. While the empirically informed model considers the same standard deviation, it samples the mean of each participant's Gaussian prior distribution from the empirical pre-stimulation measurements of Christ et al. [50] to estimate the post-stimulation results of Flögel et al. [51].

Following Bayes' rule, the posterior probability of a common cause with the given sensory input is estimated by

$$p(C=1|\Xi) = \frac{p(\Xi|C=1)p(C=1)}{p(\Xi|C=1)p(C=1) + p(\Xi|C=2)(-p(C=1))}. \tag{6.1}$$

Common cause likelihood, marginalized with respect to the physical values of X and T, is determined by

$$p(\Xi|C=1) = \iint p(\Xi|X,T)p(X,T)dXdT. \tag{6.2}$$

The likelihood of separate causes is calculated by

$$\begin{aligned} p(\Xi|C=2) = &\iint p(\chi_V, \zeta_V|X_V, T_V)p(X_V, T_V)dX_V dT_V \\ &\cdot \iint p(\chi_P, \zeta_T|X_P, T_T)p(X_P, T_T)dX_P dT_T. \end{aligned} \tag{6.3}$$

Based on the estimations of the posterior probability of a common cause provided by the different models according to (6.1), the most probable locations of the limb in question are determined by

$$\hat{X}_P = p(C=1|\Xi)\,\hat{X}_{P,C=1} + (1-p(C=1|\Xi))\,\hat{X}_{P,C=2}. \tag{6.4}$$

For substitution into (6.4), the common and separate cause position estimations $\hat{X}_{V,C=1} = \hat{X}_{V,C=1}$ as well as $\hat{X}_{V,C=2}$ and $\hat{X}_{P,C=2}$ are computed considering visual σ_V^2, proprioceptive σ_P^2, and tactile σ_T^2 sensory variances:

$$\hat{X}_{V,C=1} = \hat{X}_{P,C=1} = \frac{\frac{\chi_V}{\sigma_V^2} + \frac{\chi_P}{\sigma_P^2} + \frac{\mu_X}{\sigma_X^2}}{\frac{1}{\sigma_V^2} + \frac{1}{\sigma_P^2} + \frac{1}{\sigma_X^2}} \tag{6.5}$$

$$\hat{X}_{V,C=2} = \frac{\frac{\chi_V}{\sigma_V^2} + \frac{\mu_X}{\sigma_X^2}}{\frac{1}{\sigma_V^2} + \frac{1}{\sigma_X^2}} \tag{6.6}$$

$$\hat{X}_{P,C=2} = \frac{\frac{\chi_P}{\sigma_P^2} + \frac{\mu_X}{\sigma_X^2}}{\frac{1}{\sigma_P^2} + \frac{1}{\sigma_X^2}}. \tag{6.7}$$

Moreover, μ_X describes the currently perceived position of the foot comprising mean and standard deviation. Representing the prior information of this multisensory integration process, the computation of μ_X differs between the modeling approaches as mentioned above.

6.2.2 Results and Observations

Besides visual inspection, the informed RFI model variants are evaluated in comparison to the uniform model proposed by Samad et al. [29]. To this end, Bayes factors B_{iu} between the informed (i) and uniform (u) models are computed based on the observed data [54]. A Bayes factor exceeding 3 is assumed to show moderate evidence in favor of a model while factors above 10 would point towards strong evidence [55]. The estimates obtained by Schürmann et al. [30] from applying the different models to the data are shown in Figs. 6.3 and 6.4.

As can be seen from Fig. 6.3, the position estimates computed using the empirically informed model (grey, dashed) are close to the empirical observation of post-stimulation localization regarding mean and variance. While the conceptually

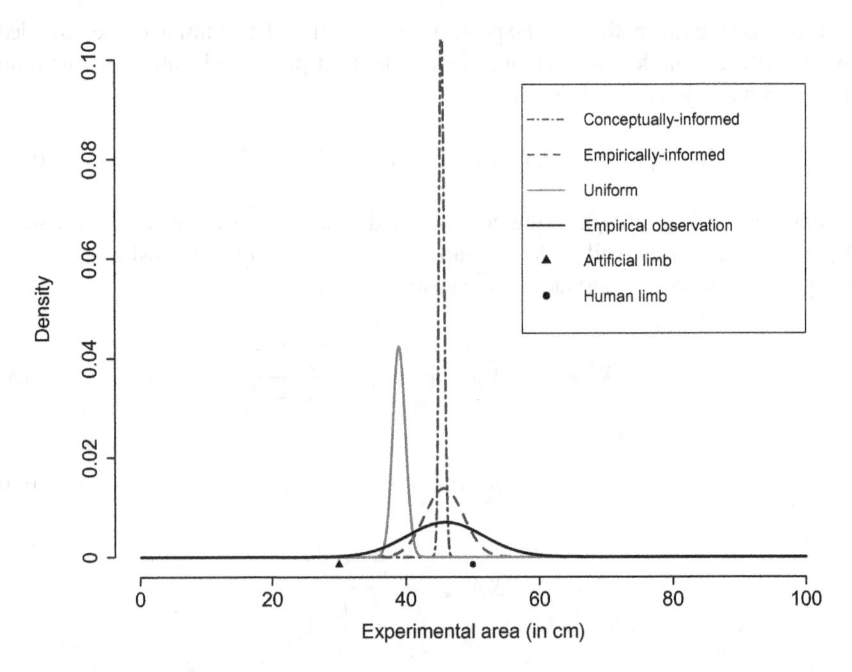

Fig. 6.3 Model predictions of position estimates after the experimental stimulation: results of uniform (light grey), conceptually informed (grey, dash-dotted), and empirically informed (grey, dashed) models compared to the empirical distribution (black). *Based on a figure from* Schürmann et al. [30]

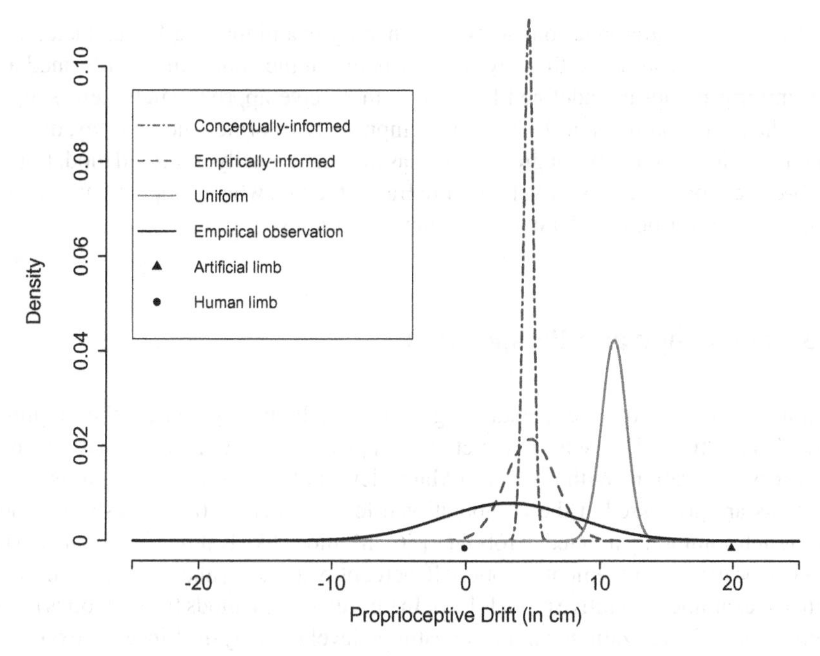

Fig. 6.4 Model predictions of proprioceptive drift: results of uniform (light grey), conceptually informed (grey, dash-dotted), and empirically informed (grey, dashed) models compared to the empirical distribution (black). Positive values indicate a drift from the position of the human limb towards the artificial limb. *Based on a figure from* Schürmann et al. [30]

informed model (grey, dash-dotted) yields a similar mean value, the resulting variance deviates distinctly from the empirically observed post-stimulation localization. For the uniform model (light grey), position estimates deviate clearly in mean and variance.

Due to this concentration of probability mass at few position estimates with the uniform model, a very large Bayes factor is observed between the empirically informed model and the uniform model. Between the conceptually informed and the uniform model produces, the Bayes factor fraction is undefined and the data do not provide support for either of the two models. The observation that the empirically informed model provides better predictions than the conceptually informed is substantiated by a positively infinite Bayes factor between both.

Going beyond post-stimulation position estimates after stimulation, Fig. 6.4 juxtaposes the prediction of proprioceptive drift by the three models (note the inverted abscissa compared to Fig. 6.3). As for the position estimates, drift results are rather accurate for both informed models, whereas the empirically informed model (grey, dashed) approximates the observed variance better than the conceptually informed one (grey, dash-dotted). The uniform model (light grey) overestimates the empirically observed mean proprioceptive drift (black) distinctly and its result do not match the observed variance as well.

Those visual interpretations are underlined by examining the Bayes factors. As for the position estimates, the Bayes factor between the conceptually informed and the uniform model is undefined because both receive approximately zero support from the data. Comparisons between the empirically informed model (grey, dashed) and the uniform model (light grey) as well as the conceptually informed model (grey, dash-dotted) both yield positively infinite Bayes factors, which supports favoring the empirically informed model over the other approaches.

6.3 Discussion and Perspectives

Aiming at an improved understanding of human body experience, the cognitive models investigated by Schürmann et al. [30] provide quantitative estimations multisensory integration. With respect to Marr's levels of analysis [9], the multisensory percepts are processed at the computational level, which distinguishes the models from such commonly applied in robotics [56]. Methods like Kalman and Bayesian filters can be used to implement robotic self-perception [18, 25] and might be combined with the examined cognitive models [30]. In this respect, methods from robotics might help to identify realizations on the algorithmic level of analysis. Since the examined cognitive models focus on multisensory integration, Bayesian approaches appear to be very suitable due to their similarities to how humans process crossmodal [4]. Still, alternative model concepts such as predictive processing [6], active inference [57], or connectionist approaches [7] might be applicable as well.

Schürmann et al. [30] transfer the Bayesian causal inference model suggested by Samad et al. [29] to the lower limbs by adapting the visual precision parameter. To improve the prediction outcome, they inform models by prior localization data and found that information from empirical pre-stimulation localization data yields the best results [30]. Besides supporting the hypothesis that participants make use of pre-stimulation knowledge, the empirically informed model is interpreted to improve estimates by considering interindividual differences. Comparisons to a conceptually informed and an uniform model reveal the empirically informed model to provide the best prediction of position estimates and proprioceptive drift. The estimates of the uniform model directly rely on the relative visual and proprioceptive precision parameters, which seems to cause its inferior prediction performance since those parameters are fixed and equal across participants [30]. Similar to the uniform model, the posterior position estimates obtained by the conceptually informed model are narrowly distributed around a specific value. Although informing the models yields a mean value that is closer to the empirical observation, the results do not reflect interindividual variance due to relying on a fixed prior across the population [30].

The limitations discussed by Schürmann et al. [30] outline directions for future developments of cognitive body experience models. Obviously, even the predictions of the favored empirically informed model do not match the empirical observation. Potentially, sensory precision parameters could be adjusted and further influencing factors might be identified in future experiments [7, 43]. Despite the remaining headroom for model improvements, Schürmann et al. [7, 30] point out that understanding bodily experience has paramount engineering potential and could strongly benefit from reliable cognitive models of human perception and cognition [7, 13, 58]. Two promising examples are the aforementioned applications in humanoids receiving human-like behavior generation and interaction capabilities as well as assistive and humanoid robotics, where devices might adapt online to their individual users' body experiences [7, 59]. Regarding the latter, extending the models to predict subjective experience based on physical measures such as the proprioceptive drift is a similarly challenging and promising endeavor.

References

1. Giummarra, M.J., Gibson, S.J., Georgiou-Karistianis, N., Bradshaw, J.L.: Mechanisms underlying embodiment, disembodiment and loss of embodiment. Neurosci. Biobehav. Rev. **32**, 143–160 (2008)
2. Christ, O., Reiner, M.: Perspectives and possible applications of the rubber hand and virtual hand illusion in non-invasive rehabilitation: technological improvements and their consequences. Neurosci. Biobehav. Rev. **44**, 33–44 (2014)
3. Deneve, S., Pouget, A.: Bayesian multisensory integration and cross-modal spatial links. J. Physiol. Paris **98**(1–3), 249–258 (2004)
4. Körding, K.P., Beierholm, U., Ma, W.J., Quartz, S., Tenenbaum, J.B., Shams, L.: Causal inference in multisensory perception. PLoS One **2**(9), e943 (2007)
5. Orbán, G., Wolpert, D.M.: Representations of uncertainty in sensorimotor control. Curr. Opin. Neurobiol. **21**(4), 629–635 (2011)
6. Clark, A.: Whatever next? predictive brains, situated agents, and the future of cognitive science. Behav. Brain Sci. **36**(3), 181–204 (2013)
7. Schürmann, T., Mohler, B.J., Peters, J., Beckerle, P.: How cognitive models of human body experience might push robotics. Front. Neurorobot. **13**, 14 (2019)
8. Sun, R.: The Cambridge Handbook of Computational Psychology. Cambridge University Press (2008)
9. Marr, D.: Vision: A Computational Investigation into the Human Representation and Processing of Visual Information. MIT Press (1982)
10. Berniker, M., Körding, K.: Bayesian approaches to sensory integration for motor control. Wiley Interdiscip. Rev.: Cogn. Sci. **2**(4), 419–428 (2011)
11. Botvinick, M., Cohen, J.: Rubber hands 'feel' touch that eyes see. Nature **391**, 756 (1998)
12. Ehrsson, H.H., Rosén, B., Stockselius, A., Ragnö, C., Köhler, P., Lundborg, G.: Upper limb amputees can be induced to experience a rubber hand as their own. Brain **131**(12), 3443–3452 (2008)
13. Beckerle, P., Salvietti, G., Unal, R., Prattichizzo, D., Rossi, S., Castellini, C., Hirche, S., Endo, S., Ben Amor, H., Ciocarlie, M., Mastrogiovanni, F., Argall, B.D., Bianchi, M.: A human-robot interaction perspective on assistive and rehabilitation robotics. Front. Neurorobot. **11**(24) (2017)
14. Longo, M.R., Schüür, F., Kammers, M.P.M., Tsakiris, M., Haggard, P.: What is embodiment? A psychometric approach. Cognition **107**, 978–998 (2008)
15. Kannape, O.A., Schwabe, L., Tadi, T., Blanke, O.: The limits of agency in walking humans. Neuropsychologia **48**(6), 1628–1636 (2010)
16. Endo, S., Fröhner, J., Music, S., Hirche, S., Beckerle, P.: Effect of external force on agency in physical human-machine interaction. Front. Hum. Neurosci. **14** (2020)

17. Roncone, A., Hoffmann, M., Pattacini, U., Metta, G.: Learning peripersonal space representation through artificial skin for avoidance and reaching with whole body surface. In: IEEE/RSJ International Conference on Intelligent Robots and Systems, pp. 3366–3373 (2015)
18. Roncone, A., Hoffmann, M., Pattacini, U., Fadiga, L., Metta, G.: Peripersonal space and margin of safety around the body: learning visuo-tactile associations in a humanoid robot with artificial skin. PloS One **11**(10), e0163,713 (2016)
19. Ulbrich, S., Ruiz de Angulo, V., Asfour, T., Torras, C., Dillmann, R.: Rapid learning of humanoid body schemas with kinematic bézier maps. In: IEEE Interational Conference on Humanoid Robotics (2009)
20. Serino, A., Bassolino, M., Farne, A., Ladavas, E.: Extended multisensory space in blind cane users. Psychol. Sci. **18**(7), 642–648 (2007)
21. Cléry, J.C., Ben Hamed, S.: Frontier of self and impact prediction. Front. Psychol. **9**, 1073 (2018)
22. Holmes, N.P., Spence, C.: The body schema and the multisensory representation(s) of peripersonal space. Cogn. Process. **5**(2), 94–105 (2004)
23. Hoffmann, M., Marques, H., Hernandez Arieta, A., Sumioka, H., Lungarella, M., Pfeifer, R.: Body schema in robotics: a review. Auton. Ment. Dev. **2**(4), 304–324 (2010)
24. Sturm, J., Plagemann, C., Burgard, W.: Body schema learning for robotic manipulators from visual self-perception. J. Physiol. **103**(3–5), 220–231 (2009)
25. Lanillos, P., Dean-Leon, E., Cheng, G.: Yielding self-perception in robots through sensorimotor contingencies. IEEE Trans. Cogn. Dev. Syst. **9**(2), 100–112 (2017)
26. Lanillos, P., Cheng, G.: Adaptive robot body learning and estimation through predictive coding. In: IEEE/RSJ International Conference on Intelligent Robots and Systems, pp. 4083–4090. IEEE (2018)
27. Nguyen-Tuong, D., Peters, J.: Model learning for robot control: a survey. Cogn. Process. **12**(4), 319–340 (2011)
28. Schillaci, G., Hafner, V.V., Lara, B.: Exploration behaviors, body representations, and simulation processes for the development of cognition in artificial agents. Front. Robot. AI **3**, 39 (2016)
29. Samad, M., Chung, A.J., Shams, L.: Perception of body ownership is driven by Bayesian sensory inference. PLoS ONE **10**(2), e0117,178 (2015)
30. Schürmann, T., Vogt, J., Christ, O., Beckerle, P.: The Bayesian causal inference model benefits from an informed prior to predict proprioceptive drift in the rubber foot illusion. Cogn. Process. **20**(4), 447–457 (2019)
31. Griffiths, T.L., Vul, E., Sanborn, A.N.: Bridging levels of analysis for probabilistic models of cognition. Curr. Dir. Psychol. Sci. **21**(4), 263–268 (2012)
32. Tenenbaum, J.B., Griffiths, T.L., Niyogi, S.: Intuitive theories as grammars for causal inference. In: Causal Learning: Psychology, Philosophy, and Computation, pp. 301–322 (2007)
33. Srivastava, N., Vul, E.: Choosing fast and slow: explaining differences between hedonic and utilitarian choices. In: CogSci (2015)
34. Thomas, M.S.C., McClelland, J.L.: Connectionist models of cognition. In: The Cambridge Handbook of Computational Psychology, pp. 23–58 (2008)
35. Quinlan, P.T.: Connectionist Models of Development: Developmental Processes in Real and Artificial Neural Networks. Taylor & Francis (2003)
36. Zhong, J.: Artificial neural models for feedback pathways for sensorimotor integration. Ph.D. thesis
37. Körding, K.P., Wolpert, D.M.: Bayesian decision theory in sensorimotor control. Trends Cogn. Sci. **10**(7), 319–326 (2006)
38. Franklin, D.W., Wolpert, D.M.: Computational mechanisms of sensorimotor control. Neuron **72**(3), 425–442 (2011)
39. Shultz, T.R., Sirois, S.: Computational models of developmental psychology (2008)
40. Lungarella, M., Metta, G., Pfeifer, R., Sandini, G.: Developmental robotics: a survey. Connect. Sci. **15**(4), 151–190 (2003)

41. Haruno, M., Wolpert, D.M., Kawato, M.: Mosaic model for sensorimotor learning and control. Neural Comput. **13**(10), 2201–2220 (2001)
42. Hoffmann, M., Lanillos, P., Jamone, L., Pitti, A., Somogyi, E.: Body representations, peripersonal space, and the self: humans, animals, robots. Front. Neurorobot. **14** (2020)
43. Beckerle, P., Castellini, C., Lenggenhager, B.: Robotic interfaces for cognitive psychology and embodiment research: a research roadmap. Wiley Interdiscip. Rev.: Cogn. Sci. **10**(2), e1486 (2019)
44. Asada, M., Hosoda, K., Kuniyoshi, Y., Ishiguro, H., Inui, T., Yoshikawa, Y., Ogino, M., Yoshida, C.: Cognitive developmental robotics: a survey. IEEE Trans. Auton. Ment. Dev. **1**(1), 12–34 (2009)
45. Hoffmann, M., Straka, Z., Farkas, I., Vavrecka, M., Metta, G.: Robotic homunculus: learning of artificial skin representation in a humanoid robot motivated by primary somatosensory cortex. IEEE Trans. Cogn. Dev. Syst. **10**(2), 163–176 (2018)
46. Martinez-Cantin, R., Lopes, M., Montesano, L.: Active body schema learning. In: Robotics: Science and Systems, Workshop on Regression in Robotics (2009)
47. Lara, B., Hafner, V.V., Ritter, C.C., Schillaci, G.: Body representations for robot ego-noise modelling and prediction. Towards the development of a sense of agency in artificial agents. In: Proceedings of the Artificial Life Conference 2016, pp. 390–397. MIT Press (2016)
48. Metta, G., Natale, L., Nori, F., Sandini, G.: Force control and reaching movements on the icub humanoid robot. In: Robotics Research, pp. 161–182. Springer (2017)
49. Körding, K.P., Wolpert, D.M.: Bayesian integration in sensorimotor learning. Nature **427**(6971), 244 (2004)
50. Christ, O., Elger, A., Schneider, K., Beckerle, P., Vogt, J., Rinderknecht, S.: Identification of haptic paths with different resolution and their effect on body scheme illusion in lower limbs. Technically Assisted Rehabilitation (2013)
51. Flögel, M., Kalveram, K.T., Christ, O., Vogt, J.: Application of the rubber hand illusion paradigm: comparison between upper and lower limbs. Psychol. Res. **80**(2), 298–306 (2015)
52. van Beers, R.J., Sittig, A.C., Denier van der Gon, J.J.: The precision of proprioceptive position sense. Exp. Brain Res. **122**(4), 367–377 (1998)
53. Hirsh, I.J., Sherrick, C.E.: Perceived order in different sense modalities. J. Exp. Psychol. **62**(5), 423 (1961)
54. Annis, J., Palmeri, T.J.: Bayesian statistical approaches to evaluating cognitive models. Wiley Interdiscip. Rev.: Cogn. Sci. **9**(2), e1458 (2017)
55. Lee, M.D., Wagenmakers, E.J.: Bayesian Cognitive Modeling: A Practical Course. Cambridge University Press (2014)
56. Siciliano, B., Khatib, O.: Springer Handbook of Robotics. Springer (2016)
57. Friston, K.J., Stephan, K.E.: Free-energy and the brain. Synthese **159**(3), 417–458 (2007)
58. Caspar, E.A., de Beir, A., Magalhães Da Saldanha da Gama, P.A., Yernaux, F., Cleeremans, A., Vanderborght, B.: New frontiers in the rubber hand experiment: when a robotic hand becomes one's own. Behav. Res. Methods **47**(3), 744–755 (2015)
59. Schürmann, T., Beckerle, P.: Personalizing human-agent interaction through cognitive models. Front. Psychol. **11**, 2299 (2020)

Part IV
Future Directions

Chapter 7
Design Considerations

Abstract Promoting technical means advances the experimental possibilities to probe human-robot body experience and improves implementing the target applications. This chapter discusses practical approaches towards wireless sensing and feedback as well as integrating psychophysiological measurement. Even with low-cost solutions, current technology holds a substantial potential to improve human-robot experience research. Moreover, an expert study exploring ways to tailor mechatronic limb designs for human-in-the-loop experiments is presented. It recommends taking inspiration from synergies to achieve high dexterity while simplifying design and control. Finally, a new hand/arm design to foster the investigation of ecologically valid scenarios through increasing wearability is discussed.

7.1 Wireless Multi-DoF Sensing and Feedback

Human-in-the-loop experiments exploring human-robot body experience require advanced interfaces facilitating dexterous control and feedback without distracting motion ranges. To this end, interfaces with wireless multiple degrees of freedom (multi-DoF) sensing and feedback improve wearability and are a promising direction.

Sensor gloves are a possible implementation and found various applications in robotics and human-robot interaction, e.g., teaching manipulation through demonstration [1–3] and rehabilitation [4], and are already applied in embodiment research [5, 6]. Since there is usually no one-to-one mapping from human to robot motions, transformations are required and often applied to overcome kinematic dissimilarities [7]. Yet, acquiring additional motion data through multi-DoF sensing should reduce this issue and yield better interaction [7, 8].

In recent years, sensor gloves with such multi-DoF sensing per finger were introduced providing similar functionality at a lower price compared to commercially available gloves [7, 9, 10]. To integrate sensing into glove-style interfaces, resistive sensors [7, 10], marker-based motion capturing [11], optical encoders [12], as well as combinations of inertial and magnetic sensors [9] are applied. Besides tracking finger motions, acquiring hand position/orientation, and a wireless interface to the host computer are relevant to increase the experimental possibilities [7]. In the

P. Beckerle, *Human-Robot Body Experience*, Springer Series on Touch and Haptic Systems, https://doi.org/10.1007/978-3-030-38688-7_7

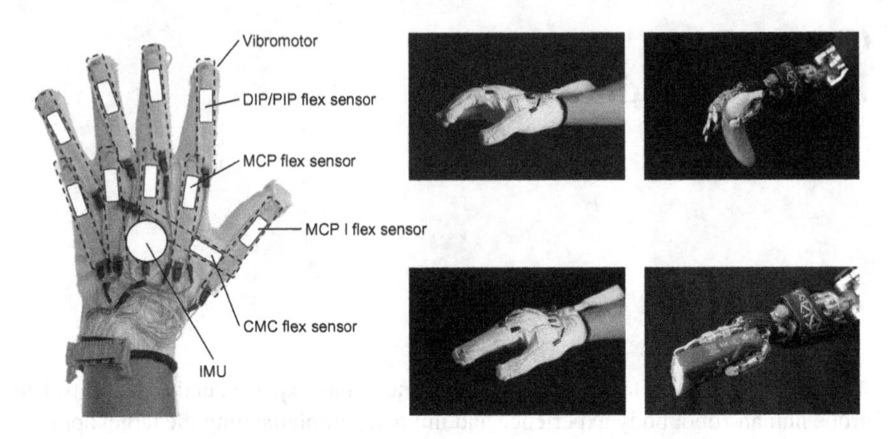

Fig. 7.1 Concept of the DAGLOVE presented in [7] (left). The vibration motor and flex sensor positions are shown for the examples of index finger and thumb. Hand motion versatility is demonstrated in teleoperated grasping (right). *Based on figures from* Weber et al. [7]

remainder, different technical aspects of the DAGLOVE by Weber et al. [7], which implements a bi-directional interface through combining wireless motion sensing with vibrotactile feedback, are discussed to identify general directions of future limb interface designs.

7.1.1 Sensor Glove Concept

Multi-DoF sensing is implemented in the DAGLOVE by measuring data from two joints per finger as shown in Fig. 7.1: interphalangeal (IP) and metacarpophalangeal (MCP) joint flexion of the index, middle, ring, and pinky fingers as well as metacarpophalangeal I (MCP I) and carpometacarpal (CMC) joint flexion of the thumb [7]. Exploiting hand biomechanics, complex motions are detected with only two flex sensors per finger since flexion and extension of the distal and proximal interphalangeal joints are coupled [13, 14]. The fixation of flex sensors needs to enable motion along the finger axis while impeding orthogonal movement, which is implemented by sewing elastic sensor pockets on the upper side of the glove [7]. Besides the flex sensors, the position and orientation of the hand are sensed by an inertial measurement unit (IMU) with a three-axis accelerometer, a three-axis gyroscope, and a three-axis magnetometer. Closing the interaction loop with the users, coin vibration motors provide vibrotactile feedback at the fingertips.

A microcontroller communicates with the host computer, connects the sensors, reads the sensor data, and commands the vibrotactile feedback using the freely available glove software [7]. Housed in an arm-shape conforming, 3D-printed electronics box, the microcontroller connects to the glove via a flexible flat ribbon cable that avoids distracting the user during operation and donning and doffing the glove [7]. For wireless operation, data is transferred via Bluetooth and power is supplied by an exchangeable and rechargeable battery pack.

7.1.2 General Applicability

From a global perspective, the conceptual idea of the DAGLOVE could be applied to interfaces for other limbs. Considering the leg, the interface orthosis applied in the RobLI experiments by Penner et al. [15] takes a similar direction by using IMUs to measure limb position and orientation and being capable of providing vibrotactile feedback. To cover more complex motions, further joints like ankle or hip might be equipped with further instrumented orthoses covering the required degrees of freedom. Increasing wearability through wireless data transfer and mobile power supply will further extend the range of possible experiments. Thereby, it can support long-term human-in-the-loop experiments in realistic interaction scenarios or even in everyday lives.

7.2 Integrating Psychophysiological Measurement

Striving to measure psychological effects objectively is very important for interpreting human-in-the-loop experiments. Considering psychophysiological data such as electrodermal activity or electromyography is very promising considering human-robot body experience research [16] and might be implemented in experimental setups [17].

An interesting example is acquiring information about the arousal of the participant through measuring electrodermal activity [18–20]. Electrodermal activity sensing includes the quantification of the general skin conductance/resistance level and phasic electrodermal responses [21]. Although responses might not always correlate with a stimulus [21], electrodermal activity might be considered as a psychophysiological measure of device embodiment [16, 22, 23].

State-of-the-art electrodermal activity measurement systems use either direct current or more complex alternate current measurements [21]. Direct current approaches provide a constant voltage to electrodes attached to the hand of the participant and evaluate the skin conductance and phasic responses through application-specific electronics and algorithms [17, 21]. For flexible application in experiments, measurement systems should be wearable and low cost, which is not the case for many commercial products [17].

Lately, first low-cost approaches systems with measurement quality comparable to commercial systems have been proposed [17, 24–26]. Among those, the system proposed by Schmidt et al. [17] increases the resolution of phasic responses through

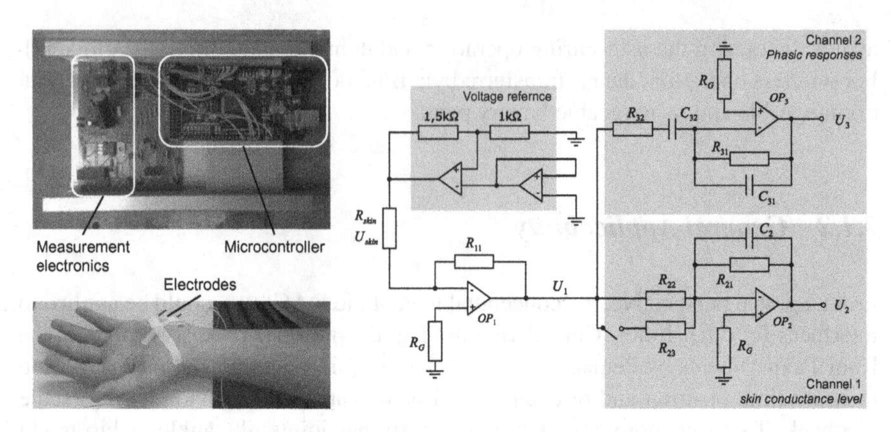

Fig. 7.2 Electrodermal measurement electronics, microcontroller (left top), placement of the electrodes on the human hand (left bottom), and circuit schematics (right). *Based on figures from* Schmidt et al. [17]

a second amplification channel based on the concept from [21]. The discussion of this concept serves as an example of integrating psychophysiological instrumentation and to derive general directions provided by that.

7.2.1 Measurement System Concept

The electrodermal activity measurement system proposed by Schmidt et al. [17] uses a low-cost approach to extend the circuit from solution based on the circuit of Boucsein [21]. The individual amplification of the skin conductance level and phasic response yields very good resolution and is shown in Fig. 7.2. A small footprint facilitates mobile use during experiments.

As shown in the schematics in Fig. 7.2, a highly stable reference voltage is applied to the skin of the participant, which is represented by the resistor R_{skin}. To distribute the low- and high-frequency contents, the signal is low-pass filtered in channel 1 and high-pass filtered in channel 2. The separate channels for amplifying the skin conductance level and the phasic responses enable covering the whole signal range despite limited A/D resolution of the microcontroller [17]. To adjust the skin conductance level measurement, resistor R_{23} can automatically be integrated in the circuit in case of lower output voltages.

7.2.2 General Applicability

Schmidt et al. [17] evaluate electrodermal data acquisition in a comparative study with a commercial product and found that data quality is appropriate for detecting external stimuli. Despite exhibiting higher noise and lower resolution than the commercial system, the proposed measurement system enables identifying and interpreting rather weak phasic responses. Hence, the system could be used to record electrodermal activity and for qualitative analysis in human-robot body experience research. Given a microcontroller with sufficient computing power, the measurement approach could easily be integrated in the sensor glove of Weber et al. [7] and improve future versions of human-in-the-loop experiments like the RobHI study by Huynh et al. [6] or the RobLI study by Penner et al. [15].

7.3 Tailoring Mechatronic Designs

Previous work has shown that mechatronic limb designs should be adapted to human-in-the-loop experiments since commercial robotic limbs usually aim at other objectives [27, 28]. Up to now, most mechatronic limb designs applied in psychological research do not reach the capabilities of the human hand [29–31] and the influence of haptic feedback is not fully understood [32–34].

Accordingly, it appears very promising to advance future human-in-the-loop experiments by tailoring mechatronic designs considering the specific requirements of such studies [35]. Through development guidelines, artificial limbs might approach the expectations of participants and perform similarly to their biological counterparts [28]. Taking inspiration from recent neuroscientific insights, low-dimensional principal control-actuation patterns, i.e., kinematic synergies [36], might be considered. Relying on synergies, the complexity of robotic hand mechanisms and controls can be reduced by exploiting inter-digit motion correlations [37–40].

One approach to systematically tailor a robotic hand design for RobHI experiments was presented by Beckerle et al. [35]. Using the Delphi approach [41], approaches and opinions of international robotic hand design and control experts were surveyed and a novel robotic hand and wrist concept was created. Subsequently, this study is presented in detail and discussed as a general approach to design experiment-specific mechatronic limbs for human-in-the-loop experiments.

7.3.1 Delphi-Based Limb Conception

Generally, the Delphi method implements several rounds of expert surveys that ask for design suggestions and opinions regarding previous-round results [41]. While responding individually, all experts thereby contribute to a collective design outcome.

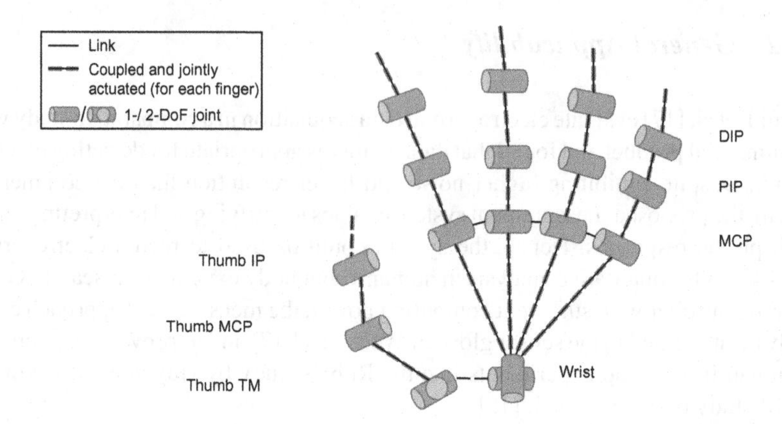

Fig. 7.3 Robot hand/wrist design concept optimized for human-in-the-loop experiments during the three Delphi rounds. *Based on a figure from* Beckerle et al. [35]

Aiming at robotic hand and wrist concepts specifically optimized for body experience research, Beckerle et al. [35] asked their experts to rank solutions for mechanisms, actuators, and control to derive an overall design concept. To this end, three successive rounds of online questionnaires were implemented and sent to experts with in-depth experience in robotic hand design and/or control. Combining fixed response and free text items, the questionnaire asked the participating experts to indicate their actuation, sensing, and control preferences and provide simplification suggestions with respect to the kinematic structure. To develop overall consensus, results were fused after each round to present improved design considerations in the subsequent round. For single-choice items, consensus was assumed to be given in case of 75% agreement and cross-checked evaluating inter-rater agreement via Kappa statistics.

The robotic hand/wrist concept developed within the three Delphi rounds is shown in Fig. 7.3 [35]. Strongly drawing from the synergy concept, the finger and thumb flexions are each implemented through kinematic coupling and joint actuation (grey dashed lines). The design requires a single actuator per finger, three actuators are required for thumb, and up to two for the wrist. Focusing on simple grasping tasks, experts were considering to actuate all four fingers with a single actuator to exploit the first synergy [42]. Experts suggested implementing elastic actuation, force/torque control, and haptic feedback [35].

7.3.2 General Applicability

Being a long-established design approach [41] that is similarly applied in other branches of science [43, 44], the Delphi method might be applied to the design of other human-in-the-loop systems. Yet, the effort of such studies is rather high and requires commitment from (external) experts, which might complicate its implementation. However, the potential outcome of Delphi studies is very informative due to the concentrated expertise. Exploiting synergies is also of fundamental value for mechatronic limb designs since it might provide valuable implications regarding kinematic simplifications for other parts of the body [45] and implementing sensory feedback [46–49].

7.4 Fostering Ecologically Valid Scenarios

To bridge the gap between experimental settings, e.g., the rubber hand illusion and its variations, and real human-robot interaction situations, more realistic and, thus, more complex scenarios should be covered in human-in-the-loop experiments [49, 50]. Fostering such ecologically valid scenarios might require for new hardware that enables users to perform a wider range of actions while investigating multisensory illusions [51]. This especially applies to free movements, which are not possible or strongly limited with current, stationary robotic limb illusion devices [5, 6, 15, 52]. While VR-based experiments approach realistic movement possibilities [53], those are limited regarding long-term stimulation and advancing technical insights [50].

Considering recent developments in wearable robotics [54–57] and integrated wireless sensing (see Sects. 7.1 and 7.2), covering more complex tasks should technically be feasible. Yet, before stepping into real applications, more complex multisensory bodily illusions might be an intermediate experimental step. In this regard, the numbness illusion is a promising paradigm demanding rather complex actions and tactile feedback [51]. This experiment studies an illusory feeling of numbness that occurs when stroking the own and another person's index fingers at the same time and combines visuomotor and visuotactile stimulation [58].

A robot-augmented version of the numbness illusion could help to go beyond previous studies and reveal fundamental mechanisms of human body experience and its influencing factors [34, 50, 51]. Robotic limbs for ecologically valid experiments are subject to complex spatial and temporal constraints [28, 51, 59]. Huynh et al. [51] designed and implemented a wearable robotic hand/arm system to elicit robotic numbness illusions (RobNI), which is discussed as an example here.

7.4.1 Wearable Hand/Arm Concept

The wearable robotic hand/arm system proposed by Huynh et al. [51] is based on the RobHI design by De Beir et al. [27] and presented in Fig. 7.4. To implement a RobNI, only index finger and thumb need to be active, but the wearable robotic hand has to allow participants' free arm motions. Accordingly, a microcontroller maps sensed human index finger and thumb flexion to the corresponding robot motions through servo motors. In case of contact at the robotic hand, contact pressures are

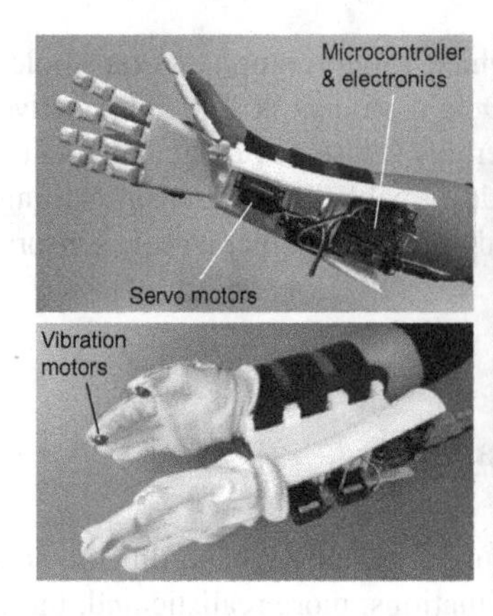

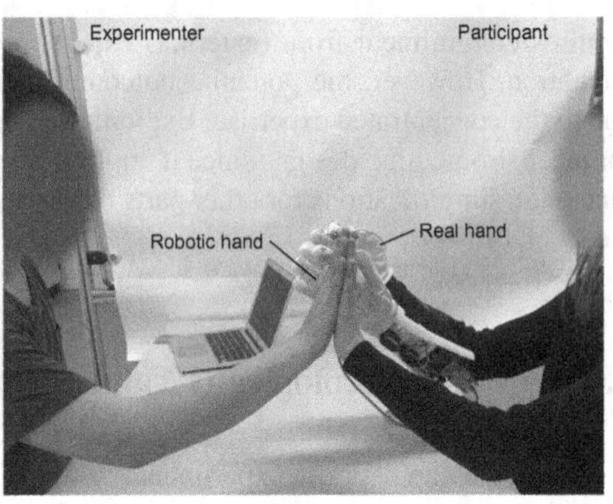

Fig. 7.4 Robot hand/arm concept from Hunyh et al. [51] extends human-in-the-loop experiments to more ecologically valid scenarios. The robotic hand wearable, weight-reduced hand/arm system includes finger actuators, vibration motors for haptic feedback, as well as electronics and microcontroller (left). The design enables investigating scenarios requiring increased motion versatility such as the robotic numbness illusion (right). *Based on figures from* Huynh et al. [51]

displayed on the human fingertips using vibration motors. For increased wearability, components have been reduced and the whole design was optimized for low weight [51].

To support bodily illusions, the hand/arm concept resembles a human hand and forearm, which additionally serves as a housing for servo motors and electronics [51]. The mechanical part consists of 3D laser sintered acrylonitrile butadiene styrene elements and steel springs implement finger joints. Through parallel attachment to the forearm of the participants via an orthosis, anatomical plausibility is ensured. Due to latencies servo motors, vibration motors, and controls, meeting temporal constraints is a tougher challenge [51]. Those system-intrinsic delays were assessed using a high-speed camera resulting in average latencies of 84.31 ms (standard deviation 12.90 ms) for the vibration motors and 110.42 ms (standard deviation 13.08 ms) for the servo motors. To achieve the latter values, finger motions had to be optimized to keep the nylon strings pretensioned and reduce the influence of backlash.

7.4.2 General Applicability

While system-intrinsic delays are crucial for all human-in-the-loop experiments, ecological validness requires wearability, which increases weight requirements and spatial constraints. The presented implementation of the RobNI is possible with low-cost components due to limitation to index finger and thumb, but more complex scenarios might require for more sophisticated solutions to achieve sufficient mechanical fidelity [60]. Yet, the study by Huynh et al. [51] pinpoints that a detailed requirement analysis considering the aspects identified by Beckerle et al. [28] fosters designing and implementing systems that facilitate examining complex multisensory illusions for all limbs. Another general insight is that small mechanical restrictions can improve global control performance, e.g., by reducing system-intrinsic delay, while assuring the required motion ranges. This could also be supported through combination with wireless interfaces and onboard power supply.

References

1. Argall, B.D., Chernova, S., Veloso, M., Browning, B.: A survey of robot learning from demonstration. Robot. Auton. Syst. **57**(5), 469–483 (2009)
2. Fischer, M., van der Smagt, P., Hirzinger, G.: Learning techniques in a dataglove based telemanipulation system for the DLR hand. In: IEEE International Conference on Robotics and Automation, vol. 2, pp. 1603–1608. IEEE (1998)
3. Fritsche, L., Unverzag, F., Peters, J., Calandra, R.: First-person tele-operation of a humanoid robot. In: 2015 IEEE-RAS 15th International Conference on Humanoid Robots (Humanoids), pp. 997–1002. IEEE (2015)
4. Prattichizzo, D., Malvezzi, M., Hussain, I., Salvietti, G.: The sixth finger: a modular extra-finger to enhance human hand capabilities. In: IEEE International Symposium on Robot and Human Interactive Communication (2014)
5. Caspar, E.A., de Beir, A., Magalhães Da Saldanha da Gama, P.A., Yernaux, F., Cleeremans, A., Vanderborght, B.: New frontiers in the rubber hand experiment: when a robotic hand becomes one's own. Behav. Res. Methods **47**(3), 744–755 (2015)
6. Huynh, T.V., Bekrater-Bodmann, R., Fröhner, J., Vogt, J., Beckerle, P.: Robotic hand illusion with tactile feedback: unravelling the relative contribution of visuotactile and visuomotor input to the representation of body parts in space. PloS One **14**(1), e0210,058 (2019)
7. Weber, P., Rueckert, E., Calandra, R., Peters, J., Beckerle, P.: A low-cost sensor glove with vibrotactile feedback and multiple finger joint and hand motion sensing for human-robot interaction. In: IEEE International Symposium on Robot and Human Interactive Communication. IEEE (2016)
8. Rueckert, E., Lioutikov, R., Calandra, R., Schmidt, M., Beckerle, P., Peters, J.: Low-cost sensor glove with force feedback for learning from demonstrations using probabilistic trajectory representations. In: ICRA 2015 Workshop Tactile & Force Sensing for Autonomous, Compliant, Intelligent Robots (2015)
9. Lisini Baldi, T., Scheggi, S., Meli, L., Mohammadi, M., Prattichizzo, D.: Gesto: a glove for enhanced sensing and touching based on inertial and magnetic sensors for hand tracking and cutaneous feedback. IEEE Trans. Hum. Mach. Syst. **47**(6), 1066–1076 (2017)
10. Lee, Y., Kim, M., Lee, Y., Kwon, J., Park, Y.L., Lee, D.: Wearable finger tracking and cutaneous haptic interface with soft sensors for multi-fingered virtual manipulation. IEEE/ASME Trans. Mechatron. **24**(1), 67–77 (2018)
11. Han, Y.: A low-cost visual motion data glove as an input device to interpret human hand gestures. IEEE Trans. Consum. Electron. **56**(2), 501–509 (2010)
12. Meng, K.L.: Development of finger-motion capturing device based on optical linear encoder. J. Rehabil. Res. Dev. **48**(1), 69 (2011)
13. Leijnse, J.N.A.L., Quesada, P.M., Spoor, C.W.: Kinematic evaluation of the finger's interphalangeal joints coupling mechanism variability, flexion-extension differences, triggers, locking swanneck deformities, anthropometric correlations. J. Biomech. **43**(12), 2381–2393 (2010)
14. Leijnse, J.N.A.L., Spoor, C.W.: Reverse engineering finger extensor apparatus morphology from measured coupled interphalangeal joint angle trajectories a generic 2D kinematic model.

Reverse engineering finger extensor apparatus morphology from measured coupled interphalangeal joint angle trajectories a generic 2d kinematic modelJ. Biomech. **45**(3), 569–578 (2012)
15. Penner, D., Abrams, A.M.H., Overath, P., Vogt, J., Beckerle, P.: Robotic leg illusion: system design and human-in-the-loop evaluation. IEEE Trans. Hum. Mach. Syst. (2019)
16. Christ, O., Reiner, M.: Perspectives and possible applications of the rubber hand and virtual hand illusion in non-invasive rehabilitation: technological improvements and their consequences. Neurosci. Biobehav. Rev. **44**, 33–44 (2014)
17. Schmidt, M., Penner, D., Burkl, A., Stojanovic, R., Schümann, T., Beckerle, P.: Implementation and evaluation of a low-cost and compact electrodermal activity measurement system. Measurement **92**, 96–102 (2016)
18. Iacono, W.G., Lykken, D.T., Peloquin, L.J., Lumry, A.E., Valentine, R.H., Tuason, V.B.: Electrodermal activity in euthymic unipolar and bipolar affective disorders: a possible marker for depression. Arch. Gen. Psychiatry **40**(5), 557–565 (1983)
19. Miwa, H., Sasahara, S.I., Matsui, T.: New mental health index based on physiological signals at transition between arousal and sleeping state. In: 2007 6th International Special Topic Conference on Information Technology Applications in Biomedicine, pp. 205–208. IEEE (2007)
20. Williams, K.M., Iacono, W.G., Remick, R.A.: Electrodermal activity among subtypes of depression. Biol. Psychiatry **20**(2), 158–162 (1985)
21. Boucsein, W.: Electrodermal Activity. Springer (2012)
22. Armel, K.C., Ramachandran, V.S.: Projecting sensations to external objects: evidence from skin conductance response. Proc. R. Soc. London. Ser. B: Biol. Sci. **270**(1523), 1499–1506 (2003)
23. Tsuji, T., Yamakawa, H., Yamashita, A., Takakusaki, K., Maeda, T., Kato, M., Oka, H., Asama, H.: Analysis of electromyography and skin conductance response during rubber hand illusion. In: 2013 IEEE Workshop on Advanced Robotics and Its Social Impacts, pp. 88–93. IEEE (2013)
24. Poh, M.Z., Swenson, N., Picard, R.: A wearable sensor for unobtrusive, long-term assessment of electrodermal activity. IEEE Trans. Biomed. Eng. **57**(5), 1243–1252 (2010)
25. Affanni, A., Chiorboli, G.: Design and characterization of a real-time, wearable, endosomatic electrodermal system. Measurement **75**, 111–121 (2015)
26. Savic, M., Gervak, G.: Metrological traceability of a system for measuring electrodermal activity. Measurement **59**, 192–197 (2015)
27. De Beir, A., Caspar, E.A., Yernaux, F., Magalhães Da Saldanha da Gama, P.A., Vanderborght, B., Cleermans, A.: Developing new Frontiers in the rubber hand illusion: design of an open source robotic hand to better understand prosthetics. In: IEEE International Symposium on Robot and Human Interactive Communication (2014)
28. Beckerle, P., De Beir, A., Schürmann, T., Caspar, E.A.: Human body schema exploration: analyzing design requirements of robotic hand and leg illusions. In: IEEE International Symposium on Robot and Human Interactive Communication (2016)
29. Padilla, M.A., Pabon, S., Frisoli, A., Sotgiu, E., Loconsole, C., Bergamasco, M.: Hand and arm ownership illusion through virtual reality physical interaction and vibrotactile stimulations. In: EuroHaptics 2010, pp. 194–199 (2010)
30. Kalckert, A., Ehrsson, H.H.: The moving rubber hand illusion revisited: comparing movements and visuotactile stimulation to induce illusory ownership. Conscious. Cogn. **26**, 117–132 (2014)
31. Ma, K., Hommel, B.: Body-ownership for actively operated non-corporeal objects. Conscious. Cogn. **36**, 75–86 (2015)
32. Hara, M., Nabae, H., Yamamoto, A., Higuchi, T.: A novel rubber hand illusion paradigm allowing active self-touch with variable force feedback controlled by a haptic device. IEEE Trans. Hum. Mach. Syst. **46**(1), 78–87 (2016)
33. Choi, W., Li, L., Satoh, S., Hachimura, K.: Multisensory integration in the virtual hand illusion with active movement. BioMed Res. Int. **2016** (2016)
34. Beckerle, P., Kõiva, R., Kirchner, E.A., Bekrater-Bodmann, R., Dosen, S., Christ, O., Abbink, D.A., Castellini, C., Lenggenhager, B.: Feel-good robotics: requirements on touch for embodiment in assistive robotics. Front. Neurorobot. **12**, 84 (2018)

35. Beckerle, P., Bianchi, M., Castellini, C., Salvietti, G.: Mechatronic designs for a robotic hand to explore human body experience and sensory-motor skills: a Delphi study. Adv. Robot. **32**(12), 670–680 (2018)
36. Santello, M., Baud-Bovy, G., Jörntell, H.: Neural bases of hand synergies. Front. Comput. Neurosci. **7**, 23 (2013)
37. Brown, C.Y., Asada, H.H.: Inter-finger coordination and postural synergies in robot hands via mechanical implementation of principal components analysis. In: IEEE/RSJ International Conference on Intelligent Robots and Systems (2007)
38. Ciocarlie, M.T., Allen, P.K.: Hand posture subspaces for dexterous robotic grasping. Int. J. Robot. Res. **28**, 851–867 (2009)
39. Liarokapis, M.V., Artemiadis, P.K., Kyriakopoulos, K.J.: Quantifying anthropomorphism of robot hands. In: IEEE International Conference on Robotics and Automation, pp. 2041–2046. IEEE (2013)
40. Santello, M., Bianchi, M., Gabiccini, M., Ricciardi, E., Salvietti, G., Prattichizzo, D., Ernst, M., Moscatelli, A., Jörntell, H., Kappers, A.M.L., Kyriakopoulos, K., Albu-Schäffer, A., Castellini, C., Bicchi, A.: Hand synergies: integration of robotics and neuroscience for understanding the control of biological and artificial hands. Phys. Life Rev. **17**, 1–23 (2016)
41. Pahl, G., Beitz, W., Feldhusen, J., Grot, K.H.: Engineering Design - A Systematic Approach. Springer (2007)
42. Santello, M., Flanders, M., Soechting, J.F.: Postural hand synergies for tool use. J. Neurosci. **18**(23), 10105–10115 (1998)
43. van der Linde, H., Hofstad, C.J., van Limbeek, J., Postema, K., Geertzen, J.H.B.: Use of the delphi technique for developing national clinical guidelines for prescription of lower-limb prostheses. J. Rehabil. Res. Dev. **42**(5), 693–704 (2005)
44. Schaffalitzky, E.M., Gallagher, P., MacLachlan, M., Wegener, S.T.: Developing consensus on important factors associated with lower limb prosthetic prescription and use. Disabil. Rehabil. **34**(24), 2085–2094 (2012)
45. Hauser, H., Neumann, G., Ijspeert, A.J., Maass, W.: Biologically inspired kinematic synergies enable linear balance control of a humanoid robot. Biol. Cybern. **104**(4–5), 235–249 (2011)
46. Bicchi, A., Gabiccini, M., Santello, M.: Modelling natural and artificial hands with synergies. Philos. Trans. R. Soc. B **366**(1581), 3153–3161 (2011)
47. Hayward, V.: Is there a 'plenhaptic' function? Philos. Trans. R. Soc. B **366**(1581), 3115–3122 (2011)
48. Bianchi, M., Serio, A.: Design and characterization of a fabric-based softness display. IEEE Trans. Haptics **8**(2), 152–163 (2015)
49. Beckerle, P., Salvietti, G., Unal, R., Prattichizzo, D., Rossi, S., Castellini, C., Hirche, S., Endo, S., Ben Amor, H., Ciocarlie, M., Mastrogiovanni, F., Argall, B.D., Bianchi, M.: A human-robot interaction perspective on assistive and rehabilitation robotics. Front. Neurorobot. **11**(24) (2017)
50. Beckerle, P., Castellini, C., Lenggenhager, B.: Robotic interfaces for cognitive psychology and embodiment research: a research roadmap. Wiley Interdiscip. Rev.: Cogn. Sci. **10**(2), e1486 (2019)
51. Huynh, T.V., Scherf, A., Bittner, A., Saetta, G., Lenggenhager, B., Beckerle, P.: Design of a wearable robotic hand to investigate multisensory illusions and the bodily self of humans (accepted). In: International Symposium on Robotics (2018)
52. Romano, R., Caffa, E., Hernandez-Arieta, A., Brugger, P., Maravita, A.: The robot hand illusion: inducing proprioceptive drift through visuo-motor congruency. Neuropsychologia **70**, 414–420 (2015)
53. Fröhner, J., Salvietti, G., Beckerle, P., Prattichizzo, D.: Can wearable haptic devices foster the embodiment of virtual limbs? IEEE Trans. Haptics (2018)
54. Dollar, A.M., Herr, H.: Lower extremity exoskeletons and active orthoses: challenges and state-of-the-art. IEEE Trans. Robot. **24**(1), 144–158 (2008)
55. Castellini, C., Artemiadis, P.K., Wininger, M., Ajoudani, A., Alimusaj, M., Bicchi, A., Caputo, B., Craelius, W., Došen, S., Englehart, K.B., Farina, D., Gijsberts, S., Godfrey, S.B., Hargrove,

L.J., Ison, M., Kuiken, T.A., Markovic, M., Pilarski, P.M., Rupp, R., Scheme, E.: Proceedings of the first workshop on peripheral machine interfaces: going beyond traditional surface electromyography. Front. Neurorobot. **5**(22), 1–17 (2014)
56. Windrich, M., Grimmer, M., Christ, O., Rinderknecht, S., Beckerle, P.: Active lower limb prosthetics: a systematic review of design issues and solutions. BioMed. Eng. OnLine **15**(3), 5–19 (2016)
57. Price, M.A., Beckerle, P., Sup, F.C.: Design optimization in lower limb prostheses: a review. IEEE Trans. Neural Syst. Rehabil. Eng. **27**(8), 1574–1588 (2019)
58. Dieguez, S., Mervier, M.R., Newby, N., Blanke, O.: Feeling numbness for someone else's finger. Curr. Biol. **19**(24), R1108–R1109 (2009)
59. Christ, O., Beckerle, P., Preller, J., Jokisch, M., Rinderknecht, S., Wojtusch, J., von Stryk, O., Vogt, J.: The rubber hand illusion: maintaining factors and a new perspective in rehabilitation and biomedical engineering? Biomed. Eng. **57**(S1), 1098–1101 (2012)
60. Nostadt, N., Abbink, D.A., Christ, O., Beckerle, P.: Embodiment, presence, and their intersections: teleoperation and beyond (submitted). ACM Trans. Hum. Robot Interact. (2020)

Chapter 8
Research Outlook

Abstract Haptic interaction plays a crucial role in achieving the embodiment of robotic devices. This chapter suggests future research directions with respect to bi-directional human-machine interfaces and considering the whole variety of the sense of touch. Aiming at robotic devices that "feel good", robots equipped with bi-directional human-machine interfaces can support neuroscientific studies to understand human cognition and identify technological challenges. To this end, a research roadmap suggests how to use technologies like haptic feedback with high spatial density and expansion, semi-autonomy, as well as intent detection. While multi-faceted tactile feedback is scarcely considered, it comprises highly relevant facets such as affective touch, social touch, or self-touch. Those kinds of feedback include non-instrumental aspects, might make a decisive contribution to device embodiment, and would benefit from technological developments of bi-directional interfaces. Discussing the related potentials, the content of this monograph is concluded and directions for cognitive modeling and human-in-the-loop experiments are discussed.

8.1 Bi-directional Interfaces

As discussed in the previous chapters and reviewed in [1], bi-directional human-machine interfaces can serve as tools to examine the human mind and its underlying neural mechanisms, which in turn supports the design of interfaces and robotic devices. Future research might aim at longer term manipulations to examine the plasticity of the bodily self [2, 3] as well as co-adaptation [1]. Paradigms might be extended from distal limbs to robotic full-body illusions considering applications such as teleoperation [1, 4, 5] using full-body interaction via exoskeletons or exosuits [1, 6].

Despite their overwhelming potentials for psychology, neuroscience, cognitive science, and clinical applications [7, 8], bi-directional interfaces still face major challenges concerning technology and the complexity of the human mind [1]. To tackle this challenge, Beckerle et al. [1] suggest a research roadmap to understand and develop bi-directional human-machine interfaces. The roadmap presented in Fig. 8.1 suggests short-term, medium-term, and long-term developments in the areas

P. Beckerle, *Human-Robot Body Experience*, Springer Series on Touch and Haptic Systems, https://doi.org/10.1007/978-3-030-38688-7_8

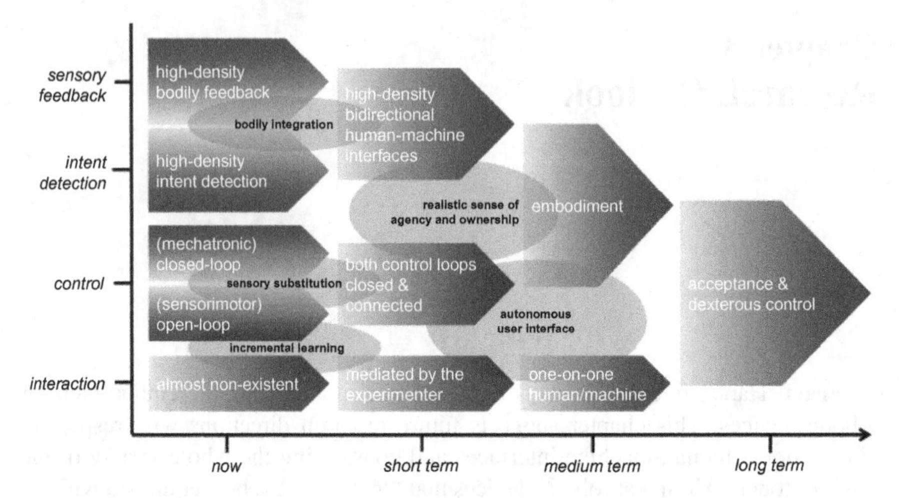

Fig. 8.1 The research roadmap suggested to understand and develop bi-directional human-machine interfaces that enable robotic experiments, which advance the investigation embodiment and embodied cognition. *Based on a figure from* Beckerle et al. [1]

of sensory feedback, intent detection, control, and interaction. It also guides merging development streams like high-density (HD) interfaces, which are not yet integrated in real applications, or fusing mechatronic and sensorimotor control loops, which are not yet interconnected, to support human-robot interaction.

As a *short-term* aim, the integration of HD tactile sensing [9, 10] and feedback [11, 12] is suggested to explore and enforce embodiment considering control delays, feedback modalities, and crossmodal processing [1]. Through fostering an interaction between the mechatronic and sensorimotor loop, e.g., via sensory substitution, bodily integration might additionally be supported [13]. Moreover, the autonomy of the machine might be modulated regarding trust of its operator and interaction could enable laboratory-controlled co-adaptation [1]. To this end, technical extensions such as those presented in Chap. 7 can advance the capabilities of human-in-the-loop experiments.

In *medium term*, the sense of agency and ownership might reach realistic levels through mutual control by human and machine [1]. Yet, psychological and physiological aspects as well as correlation, e.g., the impact of haptic stimulation on bodily experience (see Chaps. 3 and 4), will have to be investigated while connecting the control loops through the interfaces. Simultaneously, combining an autonomous user interface with mutual control is expected to yield comprehensive human-robot interaction [1], i.e., information exchange in both directions. Possible relations and interferences between semi-autonomy and embodiment need to be taken into account [13, 14], which is discussed in Chap. 4. For intuitive control sharing, haptic approaches conveying motion and interaction information through forces and torques appear very promising [13, 15].

To finally maximize user acceptance and yield dexterous control, users should be encouraged to exploit haptically enhanced co-adaptation in the *long term*. Recent works report clear learning curves in terms of performance and user experience [16–18], but interactions with embodiment are not fully understood.

8.2 Models and Experiments

Cognitive body models might be very useful to the aforementioned aims irrespective of the selected modeling approach. As outlined in Chap. 6, assistive robotic devices could learn understanding and adapting to their users' body experience while humanoid robots might generate human-like behavior from such models. As argued by Schürmann et al. [19], machine learning could be employed to achieve an integrated body representation by complementing cognitive models with established kinematic or dynamic models [20–22]. Through this, assistive devices might personalize their embodiment online and humanoid robots could improve the versatility of their behaviors and interaction capabilities on the fly [19, 23].

Human-in-the-loop approaches are a key method to inform modeling and facilitate behavior evaluation. Future research drawing on this concept will help to improve model accuracy and personalization by determining modulators and providing prior knowledge. Moreover, objective results on rather mechanistic observations like the proprioceptive drift might even be related to subjective experience or aspects of the body image. The resulting advanced cognitive models can be integrated with higher level self-perception architectures [24–26] to be applied in robot control [27, 28] or hand/tool-eye coordination [29]. Considering the research roadmap in Fig. 8.1, cognitive models can augment developments in all four areas. As software components, cognitive models could directly be integrated into intend detection, interaction control, and feedback generation and, similarly, might guide hardware development, e.g., by providing insights to sensor and actuator requirements. This could support shaping autonomous interfaces as well as sensory substitution and incremental learning to contribute to improved and individualized human-robot body experience [19, 23].

As discussed in Chaps. 3 and 4, robot-augmented and virtual human-in-the-loop experiments both have their benefits: while robotic implementation can directly guide system design, virtual reality techniques provide a wide range of experimental possibilities. Chapter 5 further shows the transferability to other parts of the body that certainly even goes beyond particular body parts [30, 31]. Irrespective of the specific experimental approach, in-depth requirement analysis is required to tune setup characteristics, which might be guided by Chap. 2. Moreover, researchers should critically consider the choice of embodiment measures, which might be psychometric, behavioral, or psychophysiological. Despite being widely used, the proprioceptive drift is subject to ongoing scientific debates since its correlation to subjective embodiment experience is not yet fully clarified [32, 33]. This is underlined by the differing observations made in the studies presented in Chaps. 3, 4, and 5. For the assessment of virtual environments and teleoperation systems, presence is a viable alternative or additional metric [34].

8.3 Variety of Human Touch

Empirical and theoretical research reviewed by Beckerle et al. [35] outlines the importance of multi-faceted tactile feedback, i.e., affective, social, and self-touch, for experiencing embodiment. While affective touch refers to low-speed tactile stimulation associated with positive feelings [35, 36], most current assistive robotic devices constrain tactile feedback regarding resolution and spatial extension focusing on the action-related touch sensations [37–40].

However, various facets of touch were found to modulate human-robot body experience [35]: self-touch can trigger illusory limb ownership [41] and affective touch can increase embodiment [42–44]. This subtle and complex influences of multi-faceted tactile information on human body experience are due to a network of receptors in the human skin that is spatially distributed and provides high-density information to the brain [35] and might call for adapting the human-in-the-loop experiment requirements discussed in Chap. 2. This might be based on user experience research considering instrumental and non-instrumental aspects [45–47]. Accordingly, application-specific feedback solutions using high-density, large-surface sensing and stimulation could improve interactive processes between touch and embodiment of assistive robotic devices [1], e.g., by enabling the transfer of emotional states [48, 49].

To this end, Beckerle et al. [35] expect an ideal human-machine interface to integrate shape-conformable, high-density, and large-surface tactile stimulation and sensing with characteristics similar to those of the human skin [50]. This might additionally be equipped with human-like sensory processing through cognitive models [19]. Despite the promising contemporary research [51], many open questions remain open and, especially, lack the consideration of affective and social touch [35]. Recent electrical and mechanical stimulation techniques [38, 52, 53] can help to explore and advance these aspects by directly activating multiple skin mechanoreceptors, e.g., with first multichannel interfaces [54, 55]. Due to the distributed, non-stationary pressure patterns of social information, high-density stimulation and sensing covering large areas of the human body appears necessary for realistic emulation [35]. To this end, stimulators and sensors need to be hardware integrated in a shape-conformable matrix aligning to the human body. Fortunately, the research of flexible sensors for distributed tactile measurement recently progresses well [10, 51, 56], which makes complex touch interaction become tangible. Future sensing systems might also implement sensors for thermal sensations associated with touch [57, 58]. Finally, holistic software integration will be necessary to process large amounts of tactile data and appropriately map between sensing and stimulation [12, 35].

Considering future applications of such interface technologies shows that suitability of different facets of touch varies depending on the domain: private-use devices like prostheses might benefit from affective tactile feedback, e.g., through improved embodiment or social communication, but professional-use systems might "only" demand for natural feedback facilitating intuitive interaction, e.g., teleoperation in hazardous areas [35]. With respect to the latter, *active touch* information that becomes available in modern robots [59] should be provided to the human user. Beyond its use in control, feeding back this data could enhance agency experience of the operator and thereby increase the intuitivity of control. Similarly, active touch feedback could help the users of assistive exoskeletons and prostheses by re-establishing the missing tactile sensations [60, 61]. Providing also *passive touch* information (see Sect. 2.2), which could automatically be available with next-generation interfaces, is promising in various assistive scenarios such as telerobotics, e.g., in case of limited vision, or prostheses, e.g., when being touched in the dark [35]. Taking the next step by displaying *affective and social touch* could support following the research roadmap given in Fig. 8.1 and enable users to integrate the robotic devices into their bodily self-representations [42–44] and improving interpersonal interaction capabilities [62, 63], e.g., by referred feedback on the residual limb of prostheses. A further step might be considering *self-touch* due to its relevance in establishing and maintaining a body representation [64] including (device) embodiment [41, 65, 66].

In the long term, we can expect such versatile and realistic tactile feedback to enhance the usability and user experience of assistive robots by fostering the development of human-machine interfaces' skin-like interaction capabilities [35]. Although this path will need further research in tight collaboration of technical and human sciences, it promises advanced robotic touch technology that enforces a real synergy between users and robots, which "feel good".

References

1. Beckerle, P., Castellini, C., Lenggenhager, B.: Robotic interfaces for cognitive psychology and embodiment research: a research roadmap. Wiley Interdiscip. Rev.: Cogn. Sci. **10**(2), e1486 (2019)
2. Apps, M.A., Tsakiris, M.: The free-energy self: a predictive coding account of self-recognition. Neurosci. Biobehav. Rev. **41**, 85–97 (2014)
3. Blanke, O.: Multisensory brain mechanisms of bodily self-consciousness. Nat. Rev. Neurosci. **13**, 556–571 (2012)
4. Blanke, O., Metzinger, T.: Full-body illusions and minimal phenomenal selfhood. Trends Cogn. Sci. **13**(1), 7–13 (2009)
5. Toet, A., Kuling, I.A., Krom, B.N., van Erp, J.B.F.: Toward enhanced teleoperation through embodiment. Front. Robot. AI **7**, 14 (2020)
6. Aymerich-Franch, L., Petit, D., Ganesh, G., Kheddar, A.: The second me: seeing the real body during humanoid robot embodiment produces an illusion of bi-location. Conscious. Cogn. **46**, 99–109 (2016)
7. Kappers, A.M.L., Bergmann Tiest, W.M.: Bayesian approaches to sensory integration for motor control. Wiley Interdiscip. Rev.: Cogn. Sci. **4**(4), 357–374 (2013)
8. Rognini, G., Blanke, O.: Cognetics: robotic interfaces for the conscious mind. Trends Cogn. Sci. **20**(3), 162–164 (2016)
9. Le, T.H.L., Maiolino, P., Mastrogiovanni, F., Cannata, C.: Skinning a robot: design methodologies for large-scale robot skin. IEEE Robot. Autom. Mag. **23**(4) (2016)

10. Büscher, G.H., Kõiva, R., Schürmann, C., Haschke, R., Ritter, H.J.: Flexible and stretchable fabric-based tactile sensor. Robot. Auton. Syst. **63**, 244–252 (2015)
11. Strbac, M., Belic, M., Isakovic, M., Kojic, V., Bijelic, G., Popovic, I., Radotic, M., Dosen, S., Markovic, M., Farina, D., Keller, T.: Integrated and flexible multichannel interface for electrotactile stimulation. J. Neural Eng. **13**(4), 046,014 (2016)
12. Franceschi, M., Seminara, L., Došen, S., Štrbac, M., Valle, M., Farina, D.: A system for electrotactile feedback using electronic skin and flexible matrix electrodes: experimental evaluation. IEEE Trans. Haptics **10**(2), 162–172 (2017)
13. Beckerle, P., Salvietti, G., Unal, R., Prattichizzo, D., Rossi, S., Castellini, C., Hirche, S., Endo, S., Ben Amor, H., Ciocarlie, M., Mastrogiovanni, F., Argall, B.D., Bianchi, M.: A human-robot interaction perspective on assistive and rehabilitation robotics. Front. Neurorobot. **11**(24) (2017)
14. Beckerle, P.: Commentary: proceedings of the first workshop on peripheral machine interfaces: going beyond traditional surface electromyography. Front. Neurorobot. **11:32** (2017)
15. Boessenkool, H., Abbink, D.A., Heemskerk, C.J.M., van der Helm, F.C.T., Wildenbeest, J.G.W.: A task-specific analysis of the benefit of haptic shared control during telemanipulation. IEEE Trans. Haptics **6**(1), 2–12 (2013)
16. Antuvan, C.W., Ison, M., Artemiadis, P.: Embedded human control of robots using myoelectric interfaces. IEEE Trans. Neural Syst. Rehabil. Eng. **22**(4), 820–27 (2014). https://doi.org/10.1109/TNSRE.2014.2302212
17. Hahne, J.M., Dähne, S., Hwang, H.J., Müller, K.R., Parra, L.C.: Concurrent adaptation of human and machine improves simultaneous and proportional myoelectric control. IEEE Trans. Neural Syst. Rehabil. Eng. **23**(4), 618–627 (2015). https://doi.org/10.1109/TNSRE.2015.2401134
18. Nowak, M., Bongers, R.M., van der Sluis, C.K., Castellini, C.: Introducing a novel training and assessment protocol for pattern matching in myocontrol: case-study of a trans-radial amputee. In: Proceedings of MEC - Myoelectric Control Symposium (2017)
19. Schürmann, T., Mohler, B.J., Peters, J., Beckerle, P.: How cognitive models of human body experience might push robotics. Front. Neurorobot. **13**, 14 (2019)
20. Haruno, M., Wolpert, D.M., Kawato, M.: Mosaic model for sensorimotor learning and control. Neural Comput. **13**(10), 2201–2220 (2001)
21. Nguyen-Tuong, D., Peters, J.: Model learning for robot control: a survey. Cogn. Process. **12**(4), 319–340 (2011)
22. Schillaci, G., Hafner, V.V., Lara, B.: Exploration behaviors, body representations, and simulation processes for the development of cognition in artificial agents. Front. Robot. AI **3**, 39 (2016)
23. Schürmann, T., Beckerle, P.: Personalizing human-agent interaction through cognitive models. Front. Psychol. **11**, 2299 (2020)
24. Lanillos, P., Dean-Leon, E., Cheng, G.: Yielding self-perception in robots through sensorimotor contingencies. IEEE Trans. Cogntive Dev. Syst. **9**(2), 100–112 (2017)
25. Asada, M., Hosoda, K., Kuniyoshi, Y., Ishiguro, H., Inui, T., Yoshikawa, Y., Ogino, M., Yoshida, C.: Cognitive developmental robotics: a survey. IEEE Trans. Auton. Ment. Dev. **1**(1), 12–34 (2009)
26. Morse, A.F., De Greeff, J., Belpeame, T., Cangelosi, A.: Epigenetic robotics architecture (ERA). IEEE Trans. Auton. Ment. Dev. **2**(4), 325–339 (2010)
27. Roncone, A., Hoffmann, M., Pattacini, U., Metta, G.: Learning peripersonal space representation through artificial skin for avoidance and reaching with whole body surface. In: IEEE/RSJ International Conference on Intelligent Robots and Systems, pp. 3366–3373 (2015)
28. Roncone, A., Hoffmann, M., Pattacini, U., Fadiga, L., Metta, G.: Peripersonal space and margin of safety around the body: learning visuo-tactile associations in a humanoid robot with artificial skin. PloS One **11**(10), e0163,713 (2016)
29. Ulbrich, S., Ruiz de Angulo, V., Asfour, T., Torras, C., Dillmann, R.: Rapid learning of humanoid body schemas with kinematic bézier maps. In: IEEE Interational Conference on Humanoid Robotics (2009)

30. Lenggenhager, B., Tadi, T., Metzinger, T., Blanke, O.: Video ergo sum: manipulating bodily self-consciousness. Science **317**(5841), 1096–1099 (2007)
31. Aspell, J.E., Lenggenhager, B., Blanke, O.: Keeping in touch with ones self: multisensory mechanisms of self-consciousness. PLoS ONE **4**(8), e6488 (2009)
32. Rohde, M., Di Luca, M., Ernst, M.O.: The rubber hand illusion: feeling of ownership and proprioceptive drift do not go hand in hand. PLoS ONE **6**(6) (2011)
33. Christ, O., Reiner, M.: Perspectives and possible applications of the rubber hand and virtual hand illusion in non-invasive rehabilitation: technological improvements and their consequences. Neurosci. Biobehav. Rev. **44**, 33–44 (2014)
34. Nostadt, N., Abbink, D.A., Christ, O., Beckerle, P.: Embodiment, presence, and their intersections: teleoperation and beyond (submitted). ACM Trans. Hum. Robot Interact. (2020)
35. Beckerle, P., Kõiva, R., Kirchner, E.A., Bekrater-Bodmann, R., Dosen, S., Christ, O., Abbink, D.A., Castellini, C., Lenggenhager, B.: Feel-good robotics: requirements on touch for embodiment in assistive robotics. Front. Neurorobot. **12**, 84 (2018)
36. Löken, L.S., Wessberg, J., McGlone, F., Olausson, H.: Coding of pleasant touch by unmyelinated afferents in humans. Nat. Neurosci. **12**(5), 547 (2009)
37. Antfolk, C., D'Alonzo, M., Controzzi, M., Lundborg, G., Rosen, B., Sebelius, F., Cipriani, C.: Artificial redirection of sensation from prosthetic fingers to the phantom hand map on transradial amputees: vibrotactile versus mechanotactile sensory feedback. IEEE Trans. Neural Syst. Rehabil. Eng. **21**(1), 112–120 (2013)
38. Schofield, J.S., Evans, K.R., Carey, J.P., Hebert, J.S.: Applications of sensory feedback in motorized upper extremity prosthesis: a review. Expert. Rev. Med. Devices **11**(5), 499–511 (2014)
39. Svensson, P., Wijk, U., Björkman, A., Antfolk, C.: A review of invasive and non-invasive sensory feedback in upper limb prostheses. Expert. Rev. Med. Devices **14**(6), 439–447 (2017)
40. Stephens-Fripp, B., Alici, G., Mutlu, R.: A review of non-invasive sensory feedback methods for transradial prosthetic hands. IEEE Access **6**, 6878–6899 (2018)
41. Hara, M., Pozeg, P., Rognini, G., Higuchi, T., Fukuhara, K., Yamamoto, A., Higuchi, T., Blanke, O., Salomon, R.: Voluntary self-touch increases body ownership. Front. Psychol. **6**, 1509 (2015)
42. Crucianelli, L., Metcalf, N.K., Fotopoulou, A., Jenkinson, P.M.: Bodily pleasure matters: velocity of touch modulates body ownership during the rubber hand illusion. Front. Psychol. **4**, 703 (2013)
43. Crucianelli, L., Krahé, C., Jenkinson, P.M., Fotopoulou, A.K.: Interoceptive ingredients of body ownership: affective touch and cardiac awareness in the rubber hand illusion. Cortex (2017)
44. van Stralen, H.E., van Zandvoort, M.J.E., Hoppenbrouwers, S.S., Vissers, L.M.G., Kappelle, L.J., Dijkerman, H.C.: Affective touch modulates the rubber hand illusion. Cognition **131**(1), 147–158 (2014)
45. Hassenzahl, M., Tractinsky, N.: User experience-a research agenda. Behav. Inf. Technol. **25**(2), 91–97 (2006)
46. Thüring, M., Mahlke, S.: Usability, aesthetics and emotions in human-technology interaction. Int. J. Psychol. **42**(4), 253–264 (2007)
47. Mahlke, S., Thüring, M.: Studying antecedents of emotional experiences in interactive contexts. In: Proceedings of the SIGCHI Conference on Human Factors in Computing Systems, pp. 915–918 (2007)
48. Hertenstein, M.J., Keltner, D., App, B., Bulleit, B.A., Jaskolka, A.R.: Touch communicates distinct emotions. Emotion **6**(3), 528 (2006)
49. Hertenstein, M.J., Holmes, R., McCullough, M., Keltner, D.: The communication of emotion via touch. Emotion **9**(4), 566 (2009)
50. Dahiya, R.S., Metta, G., Valle, M., Sandini, G.: Tactile sensing - from humans to humanoids. IEEE Trans. Robot. **26**(1), 1–20 (2010)
51. Kim, J., Lee, M., Shim, H.J., Ghaffari, R., Cho, H.R., Son, D., Jung, Y.H., Soh, M., Choi, C., Jung, S., Chu, K., Jeon, D., Lee, S.T., Kim, J.H., Choi, S.H., Hyeon, T., Kim, D.H.: Stretchable silicon nanoribbon electronics for skin prosthesis. Nat. Commun. **5**, 5747 (2014)

52. Raspopovic, S., Capogrosso, M., Petrini, F.M., Bonizzato, M., Rigosa, J., Di Pino, G., Carpaneto, J., Controzzi, M., Boretius, T., Fernandez, E., Granata, G., Oddo, C.M., Citi, L., Ciancio, A.L., Cipriani, C., Carrozza, M.C., Jensen, W., Guglielmelli, E., Stieglitz, T., Rossini, P.M., Micera, S.: Restoring natural sensory feedback in real-time bidirectional hand prostheses. Sci. Transl. Med. **6**(222), 222ra19–222ra19 (2014)
53. Li, K., Fang, Y., Zhou, Y., Liu, H.: Non-invasive stimulation-based tactile sensation for upper-extremity prosthesis: a review. IEEE Sens. J. **17**(9), 2625–2635 (2017)
54. Strbac, M., Isakovic, M., Belic, M., Popovic, I., Simanic, I., Farina, D., Keller, T., Dosen, S.: Short- and long-term learning of feedforward control of a myoelectric prosthesis with sensory feedback by amputees. IEEE Trans. Neural Syst. Rehabil. Eng. (2017)
55. Dosen, S., Markovic, M., Strbac, M., Belic, M., Popovic, I., Kojic, C., Bijelic, G., Keller, T., Farina, D.: Multichannel electrotactile feedback with spatial and mixed coding for closed-loop control of grasping force in hand prostheses. IEEE Trans. Neural Syst. Rehabil. Eng. **25**(3), 183–195 (2017)
56. Koiva, R., Zenker, M., Schürmann, C., Haschke, R., Ritter, H.J.: A highly sensitive 3D-shaped tactile sensor. In: IEEE/ASME International Conference on Advanced Intelligent Mechatronics (2013)
57. Gallo, S., Santos-Carreras, L., Rognini, G., Hara, M., Yamamoto, A., Higuchi, T.: Towards multimodal haptics for teleoperation: design of a tactile thermal display. In: IEEE International Workshop on Advanced Motion Control (2012)
58. Pacchierotti, C., Sinclair, S., Solazzi, M., Frisoli, A., Hayward, V., Prattichizzo, D.: Wearable haptic systems for the fingertip and the hand: taxonomy, review, and perspectives. IEEE Trans. Haptics **10**(4), 580–600 (2017)
59. Aggarwal, A., Kampmann, P., Lemburg, J., Kirchner, F.: Haptic object recognition in underwater and deep-sea environments. J. Field Robot. **32**(1), 167–185 (2015)
60. Witteveen, H.J.B., Rietman, H.S., Veltink, P.H.: Vibrotactile grasping force and hand aperture feedback for myoelectric forearm prosthesis users. Prosthet. Orthot. Int. **39**(3), 204–212 (2015)
61. Shokur, S., Gallo, S., Moioli, R.C., Donati, A.R.C., Morya, E., Bleuler, H., Nicolelis, M.A.L.: Assimilation of virtual legs and perception of floor texture by complete paraplegic patients receiving artificial tactile feedback. Sci. Rep. **6**, 32,293 (2016)
62. Culbertson, H., Nunez, C.M., Israr, A., Lau, F., Abnousi, F., Okamura, A.M.: A social haptic device to create continuous lateral motion using sequential normal indentation. In: IEEE Haptics Symposium, pp. 32–39 (2018)
63. Ham, R., Cotton, L.T.: Limb Amputation: From Aetiology to Rehabilitation. Springer (2013)
64. Bremner, A.J., Spence, C.: The development of tactile perception. In: Advances in Child Development and Behavior, vol. 52, pp. 227–268. Elsevier (2017)
65. Dieguez, S., Mervier, M.R., Newby, N., Blanke, O.: Feeling numbness for someone else's finger. Curr. Biol. **19**(24), R1108–R1109 (2009)
66. Huynh, T.V., Scherf, A., Bittner, A., Saetta, G., Lenggenhager, B., Beckerle, P.: Design of a wearable robotic hand to investigate multisensory illusions and the bodily self of humans (accepted). In: International Symposium on Robotics (2018)

Printed in the United States
by Baker & Taylor Publisher Services

Printed in the United States
by Baker & Taylor Publisher Services